Joanne

What to Do about CO_2

And All Those Other Ghastly Gases

PAT PALMER

Copyright © 2022 Pat Palmer.

All rights reserved. No part of this book may be used or reproduced by any means, graphic, electronic, or mechanical, including photocopying, recording, taping or by any information storage retrieval system without the written permission of the author except in the case of brief quotations embodied in critical articles and reviews.

This book is a work of non-fiction. Unless otherwise noted, the author and the publisher make no explicit guarantees as to the accuracy of the information contained in this book and in some cases, names of people and places have been altered to protect their privacy.

Archway Publishing books may be ordered through booksellers or by contacting:

Archway Publishing
1663 Liberty Drive
Bloomington, IN 47403
www.archwaypublishing.com
844-669-3957

Because of the dynamic nature of the Internet, any web addresses or links contained in this book may have changed since publication and may no longer be valid. The views expressed in this work are solely those of the author and do not necessarily reflect the views of the publisher, and the publisher hereby disclaims any responsibility for them.

Any people depicted in stock imagery provided by Getty Images are models, and such images are being used for illustrative purposes only. Certain stock imagery © Getty Images.

ISBN: 978-1-6657-2448-7 (sc)
ISBN: 978-1-6657-2446-3 (hc)
ISBN: 978-1-6657-2447-0 (e)

Library of Congress Control Number: 2022909995

Print information available on the last page.

Archway Publishing rev. date: 08/11/2022

To all those people who believe in climate change and the Green New Deal. To those who don't believe in our pending apocalypse. Temperature rise is happening but is very small. Ice ages, global droughts, and planetary changes have been going on since the planet began—most of it before the use of fossil fuels. This book is for all the people who want to protect our environment and are willing to take that hard look at what we are doing in the name of *climate change*.

Contents

Preface ix
Introduction xi

Chapter 1   What the Blazes Is Going On? 1
Chapter 2   Truth over Lies 11
Chapter 3   More of the GND 41
Chapter 4   Whereas and the Longest Sentence in the World 61
Chapter 5   Renewables, Anyone? 85
Chapter 6   Truth about Renewables 125
Chapter 7   What Sort of World Are We Headed For?
            (And Is All Hope Lost?) 147
Chapter 8   Why Is This happening? 169

Appendix I   Wind Turbines 179
Appendix II   Calculations for Wind Turbines 183
Appendix III   MW a Turbine Produces per Year 187
Appendix IV   Forests 189
Appendix V   Energy Production 191
Appendix VI   Population, Number of households, CO_2,
               Methane, SF_6 195
FIGURES and GRAPHS 197
Endnotes 199
Index 211

Preface

Why this particular book? Why now? Now I have the time. Why? I've noticed over the last several years the hate that has erupted over climate change and so much money being spent on climate change issues. I can't believe how people blindly follow political groups and truly think we are ten years away from destruction. All over the world, these political groups have convinced billions of people we are in an apocalyptic state. Massive extinction is at hand.

I felt there was a need to present the facts of my research. There are many calculations to prove my data. I spent over a year collecting information to back up my data. It is time I present it to you, the reader. With a little humor, I try to explain as simply as possible what all my research means. Have a pleasant read.

Thank you for taking the time to read this book.

Introduction

What is the Green New Deal? What is it really trying to accomplish? For years, there has been a push to save our environment using renewable energy. Now there is a politically motivated resolution that appears to have coalesced around these ideas. Many people believe that getting rid of fossil fuels will save us.

I recommend that everyone read the Green New Deal. You will find that it's not only about climate change. There are two agendas. Who'd a thunk? One agenda is the rising temperatures, which we are led to believe are sending us into oblivion. The second agenda includes purely social issues, which the proponents of this resolution are trying to link to climate change. I will define climate change.

I will be dealing with the climate issue in the Green New Deal and speaking about the social issues only as they pertain to climate change. The book goes down each point of the resolution, section by section, so the reader can follow along as I discuss them. The readers can find the Green New Deal on the internet. Happy reading.

1

What the Blazes Is Going On?

All this hoopla over climate change started in the twentieth century when scientists began warning ever louder that the planet was in trouble. Politicians soon felt the pressure to do something about it. The result of that pressure became the Green New Deal, and it is nothing less than a revolution in how we tackle climate change and our economy.

What the blazes is this Green New Deal and what is it really trying to accomplish? For years, there has been a push, off and on, to save our environment by using renewable energy. And now there is a politically motivated resolution that appears to have coalesced around these ideas. But where did the idea for this Green New Deal come from, and who wrote it? We will be looking into its origins and its contents as we journey through this swamp.

We'll break down how the Green New Deal (GND) blames human activity (fossil fuels) for temperature rise and climate change and investigate its demands. They include a demand that 100 percent of our power must come from clean, renewable sources by 2030. What? That's not even feasible as long as we need fossil fuels to build our renewables. Unfortunately, renewables aren't the savior of humankind. Fossil fuels aren't the only things that spew out CO_2 and other ghastly gases.

What is the Green New Deal? Have you read it? I suggest you do read it to follow along with this book. When you do, you might be surprised that a lot of it has nothing to do with climate change. A lot of it covers social issues that have nothing to do with climate change at all. We'll talk about all this in the coming chapters, but right now, let's see how the Green New Deal was born.

On February 7, 2019, Representative Alexandria Ocasio-Cortez (Democrat, New York) and Senator Edward Markey (Democrat, Massachusetts) presented the world with the Green New Deal, otherwise known as Resolution 109.

How exactly can we define the GND? This definition comes straight from the Green Party. The GND will create a "new, sustainable economy that is environmentally sound, economically viable and socially responsible." For them clean energy doesn't include clean coal, natural gas, biomass, or nuclear power.

So what's up with this Green New Deal? Where did it come from, and what does it actually do? It seems it all started one day when more than one hundred young adults marched into the office of Nancy Pelosi, speaker of the US House of Representatives, to protest about political action on climate. The group was called the Sunrise Movement, and Representative Alexandria Ocasio-Cortez led the march. They were presenting a message that a Green New Deal was needed, and around 140 people were arrested. (And they say January 6, 2021, was bad!)

As time went by, a tight-knit group of progressives worked on this idea but didn't put any details in this GND. A think tank called the New Consensus was then put in charge of making the GND into a policy proposal but had written almost nothing about the subject matter. It was a fairly new think tank, and they knew very little about environmental policy.[1] They wanted to combine ideas from all over the place and create one huge domestic agenda, which they did. Their main source of information came from the International Panel for Climate Change (IPCC) and the Fourth National Climate Assessment.

When released, the GND was very broad with no real specifics. Supposedly, the New Consensus is going to add the details at a later date. We're still waiting. These were the people behind the plan:

- Justice Democrats (PAC)—took care of election strategy
- Sunrise Movement—youth campaign
- Representative Alexandria Ocasio-Cortez—works inside Congress

When they had their first meeting on the GND, there were many problems and lots of arguments. One issue was Representative Ocasio-Cortez's push for 100 percent renewable energy. This became a hot topic for discussion because the push for renewable energy did not include nuclear power. Many people believed this plan, if implemented, would be way too expensive and create many brownouts across the country. The more moderate liberals suggested a 100 percent clean-energy statement. That would leave the door open to include nuclear power. To make it more palatable, the GND statement ended up being "meeting 100 percent of the power demand in the United States through clean, renewable, and zero-emission energy sources."

Before continuing, do you know what *net zero* really means? It doesn't mean getting rid of all fossil fuels. It doesn't mean getting our CO_2e levels down to zero. (CO_2e is the carbon dioxide equivalent, or the carbon footprint, not just of carbon dioxide but all greenhouse gases.) It does mean that however much CO_2e we spew out, we must remove an equal amount. What goes in must come out. It's important to mention that, because the problem with this issue is that the far, Far Left wants to get rid of all fossil fuels entirely to reach net zero. Moderate liberals feel that we can reach net zero by using brand-new technology to help get rid of our CO_2e without disrupting our entire system of living, and this includes the possible use of nuclear power.

I recommend that everyone read the GND. Once you've read it, you will see that it's not only about climate change. There are two agendas here. *Who'd a thunk?* One agenda is rising temperatures with climate change, and the second agenda speaks to social issues. We are only dealing with their climate agenda, not the social issues, although they are trying to link all the social issues to climate change. It's their way of passing legislation for these social issues. Link it to climate change. Phooey!

Let's start with the first important statement of the GND. It states they are only looking at *observable* climate change within the last hundred years. That's what it says, folks. If we take them at their word, then it's not looking at anything prior to about 1919. They say observable changes in our *climate*. Only the more primitive *observed* measurements and visual changes. But that's not what they are describing. They are describing the effects of temperature rise. And we are looking at only the last hundred years. Maybe.

So why do they bring up *preindustrial* temperatures? Real preindustrial is the time period before about 1750. Here is the problem. We all recognize that as a fact. However, many scientists suggest we take 1880 as preindustrial because that's when more accurate temperature measurements started to be recorded. We'll be asking questions about this as we go along, as to which date they are talking about. This can be a problem when discussing global warming and climate change. "What's the difference between the two?" you ask. Global warming is just that. It measures the global surface temperature change. Climate change is a measurable and observable shift and alteration in weather patterns over a *long period of time*.[2] How long? Maybe 150 years, or maybe three hundred years? Scientists are very vague about this. Some people are even trying to say that climate change could take place in thirty years. I hope they don't think we're that stupid. If they can't define what they think climate change is, then they can't say we have it. Even the IPCC Fifth Assessment Report can only say that weather creates climate from several months to millions of years. Climate is a very complex science and if you can't define it properly, how are countries going to discuss it, and make sensible climate change policy.

There's confusion over climate and weather, and a lot of people sometimes mistake climate for weather and vice versa. *Climate is long-term, permanent change, whereas weather is a day-to-day or year-to-year pattern that doesn't repeat itself.* There might be a huge flood somewhere, but there may not be another one for five hundred years, and the last one was seven hundred years earlier. That does not make for climate change just because it happens in 2022. Throughout Earth's history, the climate has always been changing. You'd think that it's normally a slow process that takes thousands of years, but I'm sure there have been fluctuations on the rate that climate has changed. Today, scientists think that humans and their fossil fuels have influenced an apocalyptic change in our *climate,* and it is moving at a much faster rate. We'll be discussing this idea throughout the book, according to what the data shows. As an example, glacier melts are a result of warming temperatures. They have been melting for about ten thousand years since the last great Ice Age. But while they may be melting a little faster, this is not climate change. This is an effect of warming. And warming is not an effect of climate change. Everyone seems to yell, "Climate change!" every time there's a flood, sandstorm, or forest fire. That is not climate change. Now if these rains that caused the floods happen every year or two for the next fifty or one hundred years, maybe even two hundred years, that might be called a localized climate change. What we are looking for is a global climate change. Let's forge on. The swamp wind is beginning to blow.

Are governments really trying to save the planet? In the last few years, groups have been pushing the agenda that humankind must change its way of life *right now* or *perish* because we only have nine years left to do it (John Kerry, 2020)! That's the point of no return. Well, I'll be jiggered. I think I'm getting the vapors! The question is this: do they really care about the planet, or is there a hidden corporate agenda to make a lot of money? Only asking. We have corporate America, Wall Street, and capitalism. I'm not saying making money is bad. Everyone has a right to make money and become very rich but not off the backs of hardworking Americans or anyone else for that matter. By making people believe there is an imminent apocalypse, they can keep everyone terrified, and big money can just keep making money. However, we also have to deal with several other issues that influence weather and climate.

Have you noticed that the population is exploding, consuming more and more resources? As this continues, the world will consume more space, leaving less and less for agriculture and energy production and destroying more of our environment. No one really talks about that. This is one of those facts that politicians and scientists leave out of their speechifying when selling us on renewables. I guess talking about population growth is a real no-no.

The world's population is expected to reach 9.7 billion people by 2050, from our current number of 7.7 billion. This comes from the United Nations Department of Economic and Social Affairs.[3] Its study also showed that the world's population could reach eleven billion by the end of the century. The US population is expected to double sometime in the 2060s, and by 2100, we will have one billion people. That's not counting the number of illegal, undocumented people pouring over our borders, or whatever the Left is calling them now. Anyway, that's a different subject.

The UN has something called the SDG (sustainable development goals). What is it? It's the "globally agreed targets for improving economic prosperity and social well-being while protecting the environment." To understand population growth, the SDG has to look at population changes in terms of births, deaths, fertility, population in different regions, migration, and more. Migration is now a major issue in population change from country to country. All this data is critical for the SDG to monitor "global progress toward achievement of the Sustainable Development Goals by 2030."[3] These are good things to study so we can see what can be done for the future of our planet.

It's really not in the best interests of climate change discussion when unnamed writers put out statements like, "Most of the current warming trend is extremely likely ... the result of human activity since the 1950's and is proceeding at an unprecedented rate over decades and millennia."[2] This is just meant to terrify us so it stops any further discussion. It means we have to go straight to renewables or die.

Do environmentalists even know what's going on other than "Save the environment!"? There are a lot of people with degrees in some kind of environmental science. Most of us don't have these degrees but are environmentalists in one way or another. Most of us are not radical environmentalists who chain ourselves to trees. We all want to save our ecosystems, rain forests, lakes, and rivers. Personally, I don't think most environmentalists have the scientific knowledge to *discuss* renewable energy, except the ones with degrees in science. That's one of the negative critiques that the GND has received. Those who wrote it weren't scientifically inclined.

Give us money, give us your vote, and we will take care of all your planetary needs. We'll save the world. Buy the Green New Deal, but we'll need your tax dollars to subsidize all the renewables to keep them working and to line the pockets of our friends. What a bunch of road apples. Bake a few pies and serve them up. Here is how it goes.

After scientists complete their work on climate, global temperatures, and climate change, they report to their masters at the International Panel on Climate Change (IPCC). The GND was set up based on the IPCC and other reports. The IPCC sends out its information (what it wants us to see) to all the governments to help them make decisions regarding climate change. After these spine-tingling reports, the gods of fear point their long, spindly fingers at us lowly, cowering humans, screeching something about how *we must* save ourselves by crawling under their long black cloak of control. Sorry, protection. Ugh! We should know these bureaucrats can't save us. Why? Because they are not so much interested in CO_2 emissions as in making money. More about that later. Lowering emissions is just a stepping-stone to making money.

What Is the IPCC?

> The IPCC was created to provide policymakers with regular scientific assessments on climate change, its implications and potential future risks, as well as to put forward adaptation and mitigation options. (IPCC)

Every few years, they put out new reports. It's huge, like an elephant in a flock of chicks. The Fourth National Climate Assessment is pretty much the same, with the same agenda.

There is no denying the planet has a slight rise in global surface temperature. However, no one can predict the future. There are too many variables. Anything can happen. Compare this with the *experts* on COVID-19. How many models did they throw at us in 2020/21? A plethora. Scientists can only give us models based on their studies and theories. Models change as new information becomes available, or they make it up as they go along. It's the same with climate models.

When our global leaders lecture us about global warming or climate change, they are only talking about what they believe is the *human factor*. They only talk about humans spewing out greenhouse gases (GHG), especially CO_2 and methane due to fossil fuels and cows. I'm not saying humans haven't contributed CO_2 and other GHGs, but the leaders have spit out a whole lot of scary flapdoodle over this. Some of it is accurate, and some of it is junk. The stuffed waistcoats suffer from the sins of omission, or they simply have no idea what's going on. We'll take a look at CO_2 and other gases (which, as you recall, is tagged CO_2e) as we ramble around looking for the truth. They only tell us what they want us to know. As cartoonist Ashleigh Brilliant says, "Opinions may change from time to time, but not the fact that I'm right."

Right off the bat, the GND tells us it's *all* the fault of humans and fossil fuels. They have to make us believe this! But that's not the whole truth! Politicians are treating us like jerks, telling us there is nothing compared to the creepy evil called fossil fuels. We must get rid of all of it. Talking about jerks, this reminds me of something I heard once. When a horse stops with a jerk ... the jerk usually falls off. Just a little cowboy humor.

According to a White House Initiative on Global Climate Change, our rise in temperature so far is only 1.0°F over the last century.[221] "Down to Earth Climate Change" tells us the current rise is 1.33°F (0.74°C),[222] and that's according to experts. But there are other experts that tell us it's much higher than that. Why? You know why. It's to convince us that our plight is dire, and the apocalypse is just around the corner. We must hurry with our renewables. It's kind of like the horse and the jerk. Experts can repeat all they want about that horse stopping with a jerk, that the jerk will not fall off. It doesn't matter how many times they say it, that *jerk* is gonna fall right off that horse!

It's important to understand how science works. Scientists collect data, compile the data, study the data, discuss the data, then compile all the data and present their *theories*. That's right, theories. Not facts. Facts are the data, the information they collect. How they interpret the data becomes the theories. Scientists can look at the very same data and come up with different *theories*. Remember this as you continue reading. Science isn't perfect. Now I'm not rebuking scientists at all. I love science. I have my degree in chemistry. That's just the nature of the beast. The so-called experts don't think you need to understand the scientific concepts around climate change and temperature rise, but they are very important in order to make informed decisions. They want us to blindly follow like lemmings. Don't do it. The framers of the IPCC reports and the arrogant bureaucrats are simply Chicken Littling us into thinking the sky is falling. But I digress.

There are so many things the universe can throw at us. Like right now. We are starting a solar storm that will affect our power grid for years. Even our planet is fickle. A lot of it has nothing to do with human activity. For instance, the earth has experienced many ice ages. Believe this or not. The ice caps began retreating some ten thousand years ago. Who'd a thunk! Maybe they are currently retreating a little faster, but most likely, centuries from now or even sooner, the glaciers will advance once again, and New York could be under hundreds of feet of ice. There is nothing humans can do to stop it.

Plate tectonics can affect changes in our climate. The movement of the continental plates changes the position of the land masses that can change the flow of the oceans. The ocean currents can then change the climate. Again, it's a slow process, but it will change. Eventually, humans and *huwomen* will be affected by these changes. I just couldn't help throwing that in. Maybe we are seeing a few of those changes now, or maybe not. Just a thought.

Humans have inflicted an immense amount of damage to our planet that we should not abide by. Things that have nothing to do with fossil fuels but can affect how heat is absorbed or reflected by the planet. Basic science tells us that if the ground is dark, the sun's heat is absorbed, and if it's light, the sun's heat is reflected. It only makes sense that deforestation is a huge contributor to heat absorption. Experts think that deforestation accounts for 23 percent of our man-made CO_2 emissions. Wow. Trees have a relationship with the planet. Who knew? Losing trees is a very complicated system, more than just absorption and reflection. It can affect the local weather patterns as well as distant weather. It can affect the amount of sunlight that is absorbed by the planet.[4] Are environmentalists and politicians even making any noise over this?

Besides our discussion of fossil fuels, we must also talk about what humans *are* doing to our planet like clear-cutting our forests, polluting our oceans, fishing with explosives, spewing toxic waste into our waterways, and so on. Things that no one brings to the table because *they* don't make money off them. It's like your hound dog hiding the fox in the hen house. All these processes influence our planet at some point if they're not already affecting us. Don't let the greed of big money blind you to the truth.

You can't have renewables without fossil fuels, so here we go!

2

Truth over Lies

We begin by charting the depths of the Green New Deal, starting with subparagraph 1 and continuing up to "Other Stuff," while looking at their goal of pushing an apocalypse narrative. We'll see how renewables are realistic or not.

As your tour guide, we will now begin our expedition into the darkest depths of the mighty Green New Deal (GND). We'll slog through miles of dangerous territory filled with quicksand and giant, man-eating plants. Sounds like a scary sci-fi movie, doesn't it? The way the GNDers are describing our future, it could very well be just that. I find that politics is like a penguin. There's a lot of flapping and loud noises, but in the end, they simply waddle away.

In this chapter, we will look at each point of the GND up to "Other Stuff." The bold statements like human activity being the cause of climate change and what happens in a changing climate are just that. They are statements, ideas, not facts. Mulling over these will give us a good start to understanding how the Left has hijacked the important issues of temperature rise and climate change for their own political agenda.

Having said that, the two documents mentioned in chapter 1, the IPCC and the Fourth National Climate Assessment, were used for setting up the GND (Resolution 109). OK. That's fine. The IPCC emphasizes keeping global temperatures below 1.5°C (2.7°F). But 1.5°C from what? According to them, from preindustrial temperatures.

Which is it exactly? Some scientists go with 1750, while many others use 1880. I'm inclined to go with 1880. Round and round we go; where we stops nobody knows ...

The GND talks about *transitioning* our energy, land, urban areas, infrastructure, buildings, and commercial industrial systems. They emphasize solutions heavily pushing the renewables. Solutions that aren't much discussed are carbon capture and nuclear. Although, since I started this project, I've been hearing a lot more about carbon capture. The GNDers dislike nuclear energy and don't have it in their renewables' arsenal. These topics and many more will be discussed throughout the book.

Let's dive right in and hope we don't smash our heads on the rocks. That would not be a good thing. If you haven't noticed yet, the GND pushes the apocalypse narrative and the demand for more renewables like solar and wind. OK, folks, here we go. Put your jackets on. It's going to get pretty chilly down there in the deep, dark caves of the GND.

Subparagraph 1 of the GND states right off that "human activity is the dominant cause of observed climate change over the past century," Key words here are *observed* and *past century*. That is a sly way of trying to tell us it's a *fact* that human use of fossil fuels is causing a climate change. And, according to its proponents, this change is apocalyptic. But it's a *theory* that our small rise in temperature, up to this point, has wreaked havoc on our climate. Question: how did scientists come up with this theory that it's all the fault of humans? Humans play a part in temperature rise, but it's not the entire picture. And that's one of my theories.

According to the IPCC, people *observed* the climate over the past hundred years. That only takes us back to about 1919. That's not even close to preindustrial, which in fact is pre-1750s. Many scientists want to use 1880 as preindustrial. There's a problem right there because that is flat-out wrong. No one standardized which year to go by. Maybe that's to keep us confused. That also doesn't help the science regarding climate change. The reason many scientists are choosing

1880 is because that's when better data on temperatures started to be collected. Wasn't that easy? We know that's not the real preindustrial age. They don't know what the temperatures actually were in the 1750s. Don't be flimflammed. The true facts are: (1) preindustrial is pre-1750, and we don't know the true temperatures (we can only estimate); and (2) by 1880, we were well into the industrial age.

Observing our climate over the past hundred years is a drop in the bucket in comparison to climate changes over the last five hundred or five thousand years, or even a million years, but the elites don't care about the truth. Climate change doesn't happen in a short period of time. Two things bear mentioning. There were a lot fewer people around in 1900 in America to observe the weather, and technology wasn't what it is today. Most people didn't really care about documenting climate observations; they cared about their survival. That's another opinion of mine. And there weren't as many scientists studying climate.

The world is full of magical things patiently waiting for our wits to grow sharper (Eden Philpott). OK, where did we leave off ... Oh, yeah. What is observed climate change? What are the scientific standards that scientists use to define the term *observed*? Nowhere have I found in the GND the parameters by which scientists gather and compare their data. There could be a standard that is understood among scientists, which is fine. Have you tried reading the IPCC? It's like wading through three feet of mud in a strong headwind. At another point, from 1900 to present, different types of people made different observations. We didn't have trained global observers. Different people saw things in different ways. We're talking about ordinary people looking and describing the world around them. Then, at some point, technology came along.

Today, there are thousands of scientists studying climate. They are called climatologists, meteorologists, geologists, and so on. They study rainfall, surface temperature, storms, and more. Each area of study has its own guidelines as to what they are looking for. We have much better technology today than in 1900 or even the 1960s.

Scientists gather in conferences to compare notes, then get together with other areas of study, like biologists, and see if their results all kind of agree. If they don't agree, then come the discussions and the heated debates. That's the simplified version. As time went on, observations became accurate technological measurements. We now use lasers and satellites to measure global temperatures. But in 1880, technology was primitive. It was only just beginning, so it seems logical that measurements then weren't as accurate as they are today.

The IPCC and climate assessment reports are like a swamp ... hard to wade through. There are scientific articles (not part of the IPCC) that explain *observable* effects that we can use. It's what we can physically see, like "glaciers have shrunk" (except they have been shrinking for the last ten thousand years), "ice on rivers and lakes are breaking up earlier, plant and animal ranges have shifted, and trees are flowering sooner." Just because ice is breaking up earlier and flowers are blooming earlier, it only means these are effects of global temperature rise and possibly a shift in the seasons.

The IPCC scientists believe that temperatures will "continue to rise for decades to come" (opinion) and "possibly rise 2.5–10°F over the next century" (again an opinion).[5] That's a humongous difference. Guys, let's try to narrow that temperature gap a bit. Understand, there are true observable changes. Not opinions. Not observable climate change. These are changes to our physical planet. Changes like the permafrost melting releasing methane. This is an observable change. We know that methane is released. We don't know how much without using sophisticated equipment to measure the amount of methane. Let's not confuse things. Modern-day scientists make these observations and measurements and compare this information with what they *think* happened in the last hundred years, then make their summaries and reports. Were they measuring methane one hundred years ago? Actually ... no! It's only been since the 1950s that we have been able to measure methane. As the resolution says, "observed climate change over the past century." But let's be reminded that the last century is a drop in the bucket as far as climate goes. Observations made one hundred years ago were made by a lot fewer people, without the aid of technology. If we don't know how much methane was released by the permafrost in 1919, we can't compare it with 2021 results. Before we get into more discussions, let's begin looking at the numbered sections of the Green New Deal, starting with "Human Activity."

(1) Human Activity

Our brains are being bombarded by spineless, simpering, and blind politicians with idiotic information that human activity (fossil fuels) is *the* cause of all our problems pertaining to greenhouse gas (GHG), temperature rise, climate change, and social issues. There are a lot of reports starting with this same declaration. They must really want us to believe all that. I don't disagree the temperature has risen very slightly. Many scientists have reported our temperature rise since preindustrial times as 0.8°–1.0°C (1.44°–1.8°F). I guess it depends on who you want to listen to. I'm going with my earlier temperatures of 1.0°–1.33°F (chapter 1).

What scientists measure is global surface temperature. They have devices set all over the world and satellites that record surface temperature. So, why, if we have such good, sophisticated measuring equipment, are there such huge differences in the statistics? Maybe it depends on who wants to scare us the most. Other scientists observe changes in our planet, like glaciers melting and changes in our coral reefs. These are observations of our physical planet, not climate changes. They haven't given us the smoking gun. For now, it just is. We're told to just accept it. And for God's sake, don't investigate it.

The GND is very deceptive because the framers didn't go into the why and how of anything. Everything is stated as fact. And they are biased in one direction. We are expected to believe all this codswallop when much of it is theory and conjecture. Nowhere do we see where they make the *scientific* leap between temperature rise and their ideas of what they think is climate change. As I mentioned earlier, there is a small temperature rise between 1.0°F and 1.33°F. Scientists do observe some changes in our physical planet, and those changes, they say, are probably caused by climate change. That's what the activist leaders want all of us to believe. Ugh! These guys are slicker than snot on a doorknob. That's how big money works.

Of course, there are some physical changes (coral reefs, polar ice melts, permafrost melts) that are the result of our slight rise in temperature but not all of them. Good scientists are continuing their experiments and research, and not all agree with the IPCC conclusions. The point is CO_2e has caused the temperature to rise about 1.0°F to 1.33°F, but there is no connection to a planet destroying climate change. There is weather, and there is climate. Scientific research measures weather over thirty-year periods for specific locations. Some periods are worse than others. Climate change happens when long-term shifts in the weather don't go back to a previous normal. One huge flood is not climate change.

There is a chemistry and physics involved in global warming, so let's look at how it works. Our atmosphere is basically comprised of oxygen (O_2) and nitrogen (N_2), but they can't absorb the sun's heat. That's why greenhouse gases (GHGs) are important, because they keep our planet warm. Too much, and the planet gets downright hot. Too low, and we freeze. Water vapor is the most important GHG. It accounts for 36 percent of the greenhouse effect, while CO_2 accounts for 20 percent. Other GHGs make up the difference. Without GHGs, Earth would be a frozen popsicle.

Here it is in a nutshell. Greenhouse gas molecules in the atmosphere absorb energy (infrared photons) from the sun. This causes molecules in the atmosphere to vibrate more. Molecules never sit completely still. They are constantly moving. These vibrating molecules start bumping into one another, picking up speed. The faster molecules move, the higher the temperature of the gas, and so goes the atmosphere.[6] That doesn't just go for CO_2. All the other GHGs absorb photons and react the same way. It's a love fest of molecules!

Without some basic knowledge of science and our planetary history, we can't make intelligent judgments about what is happening.

We can only guess that our elite leaders expect everyone to read the IPCC report and take it as gospel, because it presents everything as fact and is biased in one direction ... toward renewables. They know that most people aren't going to read all these reports. Most people demonstrating around the world have probably never read the IPCC report. They follow those political leaders who are screaming the loudest about a doomsday scenario so they can get elected.

They portray themselves as the only ones who can save the world and try to shame anyone with a different opinion. That's my opinion, for what it's worth. They are political, weaponizing bitches. I did read most of the IPCC report, the national climate assessment report, the Paris Accord (that went marginally well), and the Climate Summit in Spain (that did not go well at all). And none of these educate us as to the science behind their conclusions. When they do explain anything, they make it a convoluted piece of dipshit writing so we can't be sure even they understand it.

(2) Changing Climate

Sea Level Rise

According to the IPCC, here are the effects of temperature rise and how the rise causes these direct, observable effects:

- "The oceans are getting hotter, expanding and becoming more acidic. They are getting warmer because they absorb 90 percent of the extra heat in the atmosphere. This shift causes oceans to expand, contributes minutely to higher sea levels, and strips coral of their vivid colors. Meanwhile, nearly a third of carbon dioxide emissions end up in the oceans, triggering a chemistry change that makes the water more acidic, dissolving the shells of sea creatures. The oceans are almost 40 percent more acidic than they used to be." Of course, all the crap we are dumping into our oceans contributes nothing acidic?

Yeah, right! "Sometimes I wonder whether the world is being run by smart people who are putting us on, or imbeciles who really mean it" (Mark Twain).

- "Coral and shellfish are suffering. Coral reefs are highly sensitive to small changes in ocean levels and temperatures. The heat stresses the algae that nourish the corals and provide their vibrant colors. The algae then leave, and the corals eventually starve – an event known as bleaching. As coral reefs are home to many other species such as fish, their collapse would disrupt the entire ecosystem. Also, a more acidic ocean affects the normal carbon balance, meaning creatures with calcified shells, such as shellfish and coral, may not have enough calcium to grow."[7]
- Polar ice is melting, contributing to sea level rise and lower salinity.
- Permafrost is melting, releasing methane.

These are the *observed* and documented changes. These changes just presented are supported by observable data. Reminder: the GND did say observable changes. Let's play nice and give them a lot of space. Keep in mind that polar ice melts, permafrost melts, and rise in ocean levels (6–8 inches) are not climate changes. They are physical changes to the planet due to temperature rise but are not *climate* change.

Since the earth began, there have always been changes to global waters and coastlines. We will have more shifting and rattling through tectonic movements. Earth is an ever-changing, dynamic powerhouse. The GND has not linked, with hard data, that our small rise in temperature has caused a massive global climate change. It's their assumption that more GHGs have caused climate change. Of course, the big question is, do we have that massive climate change? I guess we'll see.

We're now going to go back twenty thousand years, then to about twelve thousand years ago. We will get a better picture of sea levels. Here is how that works. When the planet starts a warming trend (and we've had a bunch), glaciers will melt, causing sea levels to rise. Besides this, sea levels also rise when temperatures rise. If you remember your high school chemistry, when water is ice, the molecules are so close together they barely vibrate. When the ice melts, the molecules expand, and we have water, and the molecules vibrate faster. So, the oceans expand. Make water boil, and the molecules vibrate so hard they get thrown out as steam. The same applies to the ocean, except it doesn't boil or turn to steam.

Back about twenty thousand years ago, temperatures rose 4°C (7.2°F). The oceans rose. We can't blame that on human use of fossil fuels! After the warming ended, sea levels eventually stopped rising and settled at about 480 feet higher than what it was. The current sea levels reached their final levels about three thousand years ago. From three thousand years to about one hundred years ago, sea levels have just been rising and falling slightly. Since 1880, the temperature has risen roughly between 1.0°F and 1.33°F. Subsequently, sea levels have risen between six and eight inches.

Over the last several millennia, we have had large amounts of sea level rise.[8] The climate change obsessed want to scare everyone into believing that ocean rise has just been happening for the last hundred years. They are trying to convince us sea levels will rise so much in the next one hundred years that we'll lose all our coastal communities and infrastructure entirely. That's because they want us to believe in the doomsday scenario so our tax dollars can be thrown at political friends in the renewable business—big political donors.

Let's continue. Every huge storm, of which there are various kinds, brings death and economic loss, no matter when it occurs. We'll always have death and destruction, especially if we want to build right on the beach, tornado alley and hurricane regions. It happens to seem worse now because we take up more space. I'll just move after our house starts going underwater. Our house is 248 feet above current sea levels, so I'll head for higher ground. Maybe Las Vegas will become the new beachfront community. Perhaps. It's a theory.

Remember when Californians were told that the San Andreas

Fault was on the verge of an earth-shattering earthquake and western California was going to fall into the ocean? I've lived here for sixty-seven years, and I'm still waiting. Eventually, our land masses will change because of tectonics. It may take a million years. Our descendants will have to deal with that if they are still here and there hasn't been some extinction event. Our planet has had several of those. It makes you wonder, but let's forge on.

Increase in Wildfires

Next on the list is the increase in wildfires. This is going to be interesting because with the information given here, you will have to decide whether these wildfires are from temperature rise and climate change due to fossil fuels or just plain human carelessness. There are several reasons for the increase in the number of fires and the size of fires.

> Nearly 85 percent of wildland fires in the United States are caused by humans. Human-caused fires result from campfires left unattended, the burning of debris, equipment uses and malfunctions, negligently discarded cigarettes, and intentional acts of arson.[9]

Include in that list downed power lines and so on. Surely, you can think of more. Other reports say forest fires were 90 percent human caused, not from fossil fuels. For now, we'll go with 85 percent, which is probably a good average. Humans shouldn't be made to appear too stupid and lacking in care for their natural world. It's just that population numbers have gone way up, so more people visit and live in our forests than ever before. Thought of the day: human carelessness is the number one cause of forest fires, *not* the use of fossil fuels.

Before we get into wildfires too much, let's just say something about the terrible fires in Australia. It was extremely sad. They have the same problem we have—human disregard. According to some reports, more than two hundred people have been arrested for setting catastrophic brushfires. I've discovered that's not quite true. It's misinformation. Being more on the conservative side, I would love to hear that 183 or 200 were arrested for arson. But the math just doesn't add up. Here is what happened. New South Wales Police reported that there were around 183 people associated with "brushfire-related offenses." Only twenty-four were arrested for deliberate arson. People were cited for not complying with the total fire ban or discarding a cigarette, and so on. Many of these people were not arrested.[10] What gets me is there are people out there who will take the 183 number and spread that news because it fits their agenda. Sadly, I fear there are too many news outlets who went with this story. All that political garbage does not help the truth of the situation. Greenhouse gases and global warming are real issues we need to discuss using the truth. The quotation below sums it up pretty well:

> The predilection of media and stars to taking any tragedy and using it as a fulcrum for their own cause celeb does less to amplify the need for action ..., and more to highlight the myopia that makes the public at large roll their eyes. Extinction Rebellion devotees of teen climate activist Greta Thunberg, and simplifications of complex issues make it easy for people to dismiss climate concerns as baseless. Stating facts and offering solutions will do more to convince people of the need for change than easily dismissible hyperbole.[10]

The other 15 percent of wildfires are natural causes like lightning. We've always had thunderstorms and lightning. Nothing new there. Some years are worse than others. That's the planet at work. Let's go back to humans. Since the world population is growing, more people are enjoying the thrills of camping and driving through our beautiful natural world. More careless people, more fires. That has nothing to do with climate change or fossil fuels. The leftist activists and politicians will use the fossil fuel argument to take us down. Because humans use fossil fuels, it must cause the temperature to rise, which makes fires worse. This might not be true. We'll see.

The second part of our discussion has to do with the size of wildfires. The proponents of climate change would like us to think that wildfires are larger because of temperature rise. This is where you will have to decide for yourself what's really going on. Let's discuss what happens in the US. Other countries may or may not have similar policies. When wildfires occur, there are wildfire managers to evaluate each fire. The managers answer two important questions. First, does the fire threaten people or personal property? If yes, these fires are fought hard to put them out. If fires move away from populated areas, then sometimes they are allowed to burn. Firefighters focus on the threat to human lives.

The second question: is the fire in an area like national parks or forests far away from human activity? Is access to the fire good or bad? These fires may be due to natural causes like lightning, which activists will tell you is due to climate change. Ugh. The fires may be allowed to burn, sometimes turning them into giant infernos. In parts of America, we have huge, very tall trees. In California, the idea is that these fires burn the underbrush, so the ecosystem gets a fresh start.[9] Environmentalists didn't want forest management; they wanted the forests to be left alone to burn naturally. Supposedly, it's OK to maim and kill thousands or even millions of forest animals. The fire management teams do the best they can to control all fires. They just can't do it all or threaten the lives of firefighters and the public. This is not climate change.

While we are talking about fires, remember burning wood releases greenhouse gases. The two worst are carbon dioxide and oxides of nitrogen, like nitrous oxide. Nitrous oxide is acidic and, in the atmosphere, makes acid rain. In 2018, there were 58,083 wildfires in the US, which burned about 8.8 million acres.[11] Maybe we can determine approximately how much wood burns per acre, then calculate how much CO_2 is released. Approximately. According to

Forest2Market, the amount of wood per acre varies a lot. We're only talking about timber.

The average amount of timber for clear-cutting (cutting all the trees) averages eighty-seven tons per acre. If you just thin the timber, the average amount is thirty-two tons per acre. But this all depends on whether the forest is wild or well managed. Then there is plantation clear cut, which averages ninety-nine tons per acre.[12] That's pretty much the same as clear-cutting a forest. So, eighty-seven tons per acre is a good average. Before we can make any calculations, we must have all our mathematical units in order.

See Appendix IV, "Forest," for the calculations. We'll just cover the results here. Not everyone likes to wade through calculations.

For every ton of timber burned, 1.9 tons of CO_2 are released.

Fact: 8.8 million acres burned in 2018 in the US.[11]

Average ton per acre is 87.

In 2018, 765.6 million tons of timber burned.

One and a half billion tons of CO_2 were released from timber burned in wildfires ... approximately.

That's just in the US. And don't forget about all the other gases like N_2O. So, why do we just let our forests burn? Good question. It's one that I don't know the answer to.

Now, getting back to my original argument of humans being responsible for 85 percent of wildfires. That has nothing to do with fossil fuels. It has everything to do with carelessness and simply not understanding your environment. That goes for world leaders as well. Sometimes it would be better if they kept their mouths shut and looked stupid rather than open them and prove it.

It's worth mentioning how these fires are managed. California's Kincaid Fire came very close to some homes. There were high winds for several days. These are Santa Anna winds that regularly hit the state. They are nothing new to California. They would have known this, so the powers that be should have taken the opportunity to vigorously fight this fire from the beginning to keep it from spreading.

It seems that never happened, but I'm only going by a news article. They could see the fire and smoke over the hills getting worse. The news crews weren't allowed up there, so they couldn't see that the fire wasn't being fought while there was still time before the winds came. This was just one observer's opinion. I thought it was worth mentioning.[14] Don't shoot me; I'm just the messenger.

My opinion: the number and sizes of these wildfires are not due to climate change. Did I just commit a sin? Of the 58,083 fires in 2018, 85 percent were caused by human carelessness and not the use of fossil fuels. If we take that away, we are left with 8,713 fires that were caused by nature itself. So, even with highly caring humans, we will still have some accidental fires. That's only to be expected.

You can only blame wildfires on human carelessness! Ugh! I hear the Dixie Fire in California was caused by electrical lines.

Here are the last facts on burning wood. There are people using wood for campfires, fireplaces, and burning debris. I haven't even accounted for this in my statistics. Worldwide, about one-third of the world's population is using wood or biomass for cooking and heating.[15] They have no other options. That's all I have to say about wood, for now.

Oh yes, don't forget about our increasing population. More people will be tramping in and around our forests, causing more damage and more fires. More people ... more fires.

Severe Storms

Hurricanes

We have always had hurricanes. We'll continue to have them. They do damage and cause death. Blaming them on climate change is like blaming a baby for crapping in its diaper. It's going to happen. Besides, one hundred years is a very short time frame to make any comparisons. And there was no way to really measure hurricanes before the 1970s. We can only estimate them from writings and descriptions. We should be less interested in the whining of the elites because their precious Miami and Malibu are going to be under water. That's what happens when you live right on the water's edge. Build on low ground, like the sand, and you are bound to have trouble.

Today's best scientific method for measuring hurricanes is called the Saffir-Simpson Hurricane Wind Scale. But it was only developed in the late 1960s. How then could they measure storms accurately before that? you ask. Well, the techniques can't compare with our current methods. Everything before 1970 was not based on one specific criterion, as it was after 1970. They studied the storms, flew around the storms, used the scientific process of the time, and made detailed descriptions, which helped make for better tracking. Back in the early days, around 1910, they used landline telegraphs and ships to send observational reports that a hurricane was on the way. Then around 1943, Joseph Duckworth flew his single-engine airplane into a hurricane. After that, airplane reconnaissance was the cat's meow. But they still could not measure a storm. As time went on, measurements got better and better.[16] I realize we're only supposed to go back a century, but I think going further back might be more valuable.

I'm not going to list every storm in the last hundred years, but I will explain a few. The really bad ones. Working backward, we've had hurricanes from 2021 back to 2015. The year 2015 was a bad year. There was Tropical Cyclone Patricia, category 5, with measured wind speeds of 221 mph. It landed in Mexico as a category 4 with 150 mph winds. It was the strongest measured tropical cyclone ever recorded.[17] We haven't had one since. What? you say. How about Katrina? She was a category 5 but hit land as a category 3 at 120 mph. She was so bad because she was very wide and hit New Orleans head-on, which caused a tremendous amount of death and destruction and massive flooding. That's not climate change. That's just what happens sometimes. New Orleans is a low-lying city, and major flooding often occurs. That's a problem all along the Gulf Coast. She became big news for political reasons. Help didn't get there fast enough from the government. It was a very sad situation.

Below is a graph of hurricanes going back to 1851. But I want to go back even further to the Great Hurricane of 1780. It hit at the end of the American Revolution. The hurricane was part of a horrible year for storm activity. Scientists have estimated it had wind speeds around 200 mph, although they are still discussing if that's correct or not. In any case, it was big. I think there were two or more category 4s that year.[18] The graph below shows storms along the American coastline. It's not everything, but I don't want to get it too complicated.

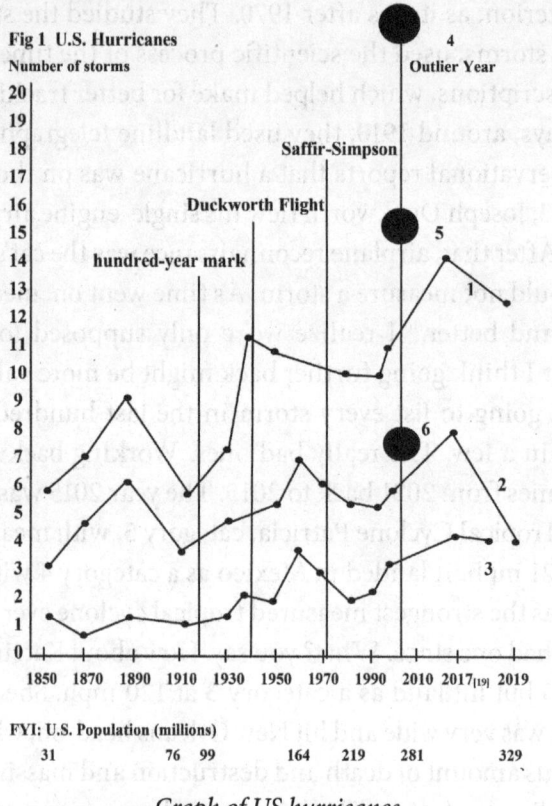

Graph of US hurricanes

Legend: dots in graph are averages for every ten years.
Named tropical storms—*1*
Hurricanes categories 1 and 2—*2*
Major hurricanes, categories 3, 4, and 5—*3*

1943—first airplane flight by Joseph Duckworth
1970s—Saffir-Simpson method established
Outlier year for hurricanes:
Tropical Cyclone Patricia 2015 Category 5
Named storms (twenty-eight)—**4**
Categories 1 and 2 (fifteen)—**5**
Major hurricanes, categories 3, 4, and 5—**6**

After reading all the research, prior to 1943, when airplanes were a giant leap in the tool chest, the ability to find, track, and measure hurricanes was hit-and-miss at best. It relied on ships and visual recognition to identify storms. Just because there were years of no major storms, this could have meant no one was around to see or record them. There were probably storms out there. To end this series on hurricanes, the only good data comes after the 1970s. Before the 1930s, it's not very reliable data. Between the 1930s and 1970s, it's only somewhat reliable. As technology gets better, the numbers go up, but we'll have to track storms for another fifty years to see any reliable trend. There isn't enough reliable data in the last hundred years that the GND professes to use. Remember, the population is continuing to increase as time goes by. More people means more death and destruction but better tracking.

Tornadoes

Meteorological events come in many forms, but one of the most violent and unpredictable is the tornado. We have the same problem with tornadoes as we did with hurricanes. Early on, before the introduction of high-tech detection and measurements, tornadoes could only be observed. There were several problems in the early years. The country was not well populated, so even a large tornado could have gone undetected or recorded. The number of tornadoes would have been underestimated and with less damage due to fewer and smaller cities, towns, and farms.[20]

Today we use the Fujita Scale (F-scale). It rates the intensity.

F-0   <73 mph   F-3 158–206 mph
F-1 73–112 mph   F-4 207–260 mph
F-2 113–157 mph   F-5 261–318 mph[21]

The scale was introduced in 1971 by Ted Fujita along with Allen Pearson. It was then updated in 1973, taking information on the path, length, and width. This is the Enhanced Fujita Scale. Using these guidelines, tornadoes were retroactively rated back to 1880 (at least the ones they knew about).[21] When the country had a smaller population, not all tornadoes were observed, reported, and counted. Below is the graph of the number of tornadoes versus time.

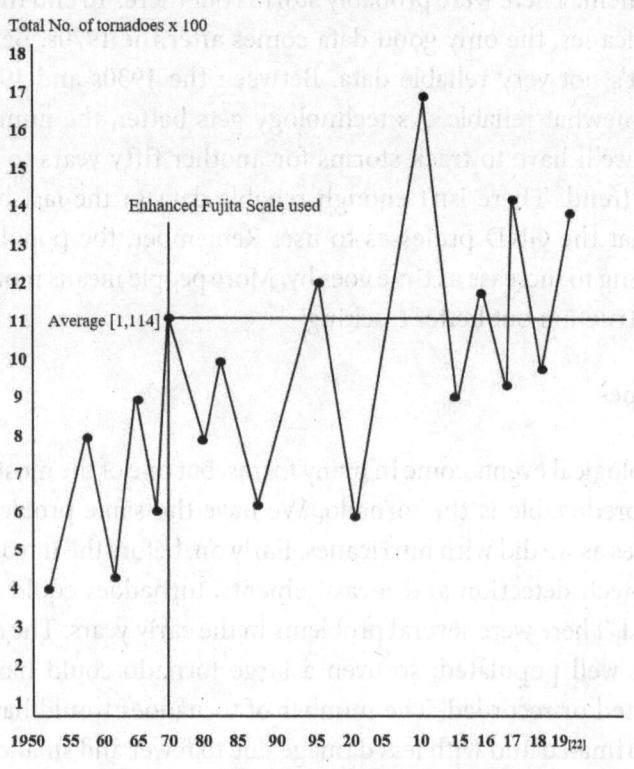

Fig 2 Tornadoes. Annual number tornadoes in the U.S. 1953-2019

Graph of tornadoes

Fig 3 Tornadoes by Year 1953-2019

Year	No.	Year	No.
1952*	422	1983	931
1954	550	1984	907
1955	593	1985	684
1956	504	1986	765
1957*	858	1987*	656
1958	564	1988	702
1959	604	1989	856
1960	616	1990	1133
1961	697	1991	1132
1962	657	1992	1297
1963*	463	1993	1173
1964	704	1994	1082
1965*	897	1995*	1234
1966	585	1996	1173
1967	926	1997	1148
1968	660	1998	1424
1969*	608	1999	1342
1970	653	2000	1071
1971	889	2001	1214
1972	741	2002*	624
1973*	1102	2005	1265
1974	945	2008	1629
1975	919	2010	1282
1976	834	2011*	1703
1977	882	2013*	903
1978	789	2015*	1178
1979	855	2016*	976
1980*	866	2017*	1418
1981	782	2018*	991
1982	1047	2019*	1407

Points plotted on the graph *
Enhanced Fujita Scale in use ———

Tornadoes by year

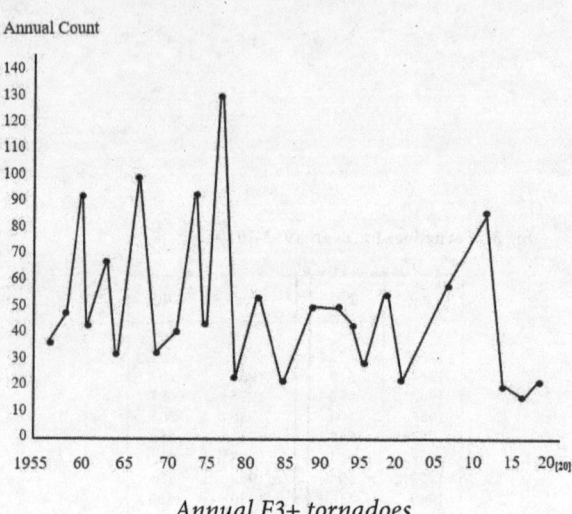

Fig 4 Annual U.S. number of F3+ tornadoes

Annual F3+ tornadoes

The tornadoes in the graph would probably have been reported using pulse-Doppler radar, which was developed in 1958. This resulted in more tornado reports. Figure 4 shows there has been very little trend in the stronger tornadoes in the past sixty-five years.[20]

Going back into the historical records, we can see there were fewer observed storms of all kinds before the mid-twentieth century from all the graphs. We can also see that when technology improved and was put into use and the population grew, the number of storms went up. Population always works itself into the mix, doesn't it?

Severe Snowstorms and Blizzards

There have been major winter storms and blizzards in our known history: March 1888, January 1922, February 1978, March 1993, January 1996, February 2010, December 2010, and so on.[23] Today, we have a lot more information. We can check the weather instantly on our phones. It comes from our satellites, which tell us just about everything we need to know about our weather. They also help to predict future storms. They are really quite handy.

The first satellite, TIROS-1, was launched on April 1, 1960.[24]

The day the first pictures came through must have been incredibly exciting, like finding a buffalo in your living room. Since then, we have launched many more satellites, each with better technology, giving better and better data. That's why we appear to have more storms and a better accounting of our storm systems. And don't forget, as the population grew, the observations got better.

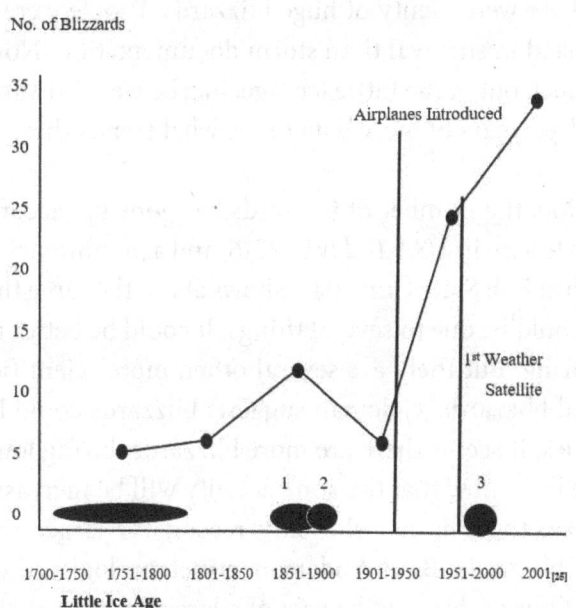

Blizzards versus time

Legend:

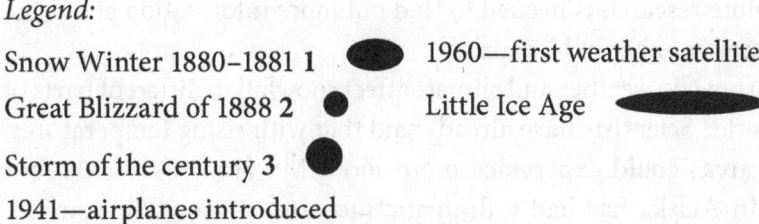

Snow Winter 1880–1881 **1**
Great Blizzard of 1888 **2**
Storm of the century **3**
1941—airplanes introduced
1960—first weather satellite
Little Ice Age

There are no easy answers when it comes to snow. Scientists are not positive when it comes to snow because it changes from place to place and year to year. We are also coming out of our Little Ice Age. It's a conundrum, to say the least, trying to get answers about the weather in terms of climate change. With satellite technology, we can see decreasing snow cover. You'd think it's obvious that global warming is the cause. I don't think it's that simple.[26] We don't really know much about blizzards from 1700 to 1800. There was a mini ice age. I imagine there were plenty of huge blizzards. People were probably more interested in survival than storm documentation. Now that we are pretty much out of the Little Ice Age, maybe we need another fifty to one hundred years of more data to see what trends there will be in the future.

Since 1960, the number of blizzards has gone up, according to a piece by Doyle Rice in *USA Today* in 2016, and a preliminary report by Jill Coleman at Ball State University shows about the same thing since 1960. This could be due to several things. It could be better reporting and monitoring. But there are several other, more scientific reasons for increased blizzards. Coleman suggests blizzards could be tied to sunspot cycles. It seems there are more blizzards during low sunspot activity. We'll see now that the sun's activity will be increasing.

Sunspot activity is not the only reason for larger or smaller numbers of blizzards. Brad Anderson, a meteorologist, thought the frequency of blizzards could be part of a larger picture of changes in "large-scale climate patterns" and that these climate patterns could be "the bigger driver of blizzard cycles." However, Coleman did say that more research is needed to find out more information about the increased number of blizzards.[27]

So how do weather and climate affect snowfall in different parts of the world? Scientists have already said that with rising temperatures, some areas could experience more snow. Mt. Hunter's 14,000-foot peak in Alaska has had a dramatic increase in snowfall. Summer snow has risen 49 percent since the mid-nineteenth century. There are reasons for that, and it has to do with warm tropical air flowing to the state, creating low-pressure systems that love making snow. Scientists are still in conflict as to exactly how and why global warming causes more moisture in the air.

Here is my opinion. There are a whole lot of ideas about snow and blizzards. They are all related one way or another. Our world is in continuous flux. Global warming, as small as it is right now (1.0°F to 1.33°F), could be causing some changes in our snowfall but not much. I'm not an expert, but I think scientists are still trying to understand the mechanisms of how and why all this happens. You can't just say that a certain type of weather is caused by global warming, because next year and the year after might surprise you.

Droughts

Droughts are a fickle thing. They can last a few years or many years. Do we talk about all our droughts or just a few, like in California? We were told the drought in California would continue for many years to come. Recently, it seems the drought may be broken. Now, in 2021, we are back in a severe drought event. The global activists, including many scientists, don't know what is coming. They are flapping their gums because they are the almighty—because the IPCC wants the *right answers*, their answers. Be banished if you don't agree with them! According to the left political elites, since the temperature is rising, it's going to continue to rise, and no one but they can stop it, so we must go to their renewables. At least that's what they say. That's their plan. If we don't align with them, the temperature will continue to rise until we burn up. A bleak future, isn't it?

Something I just read shows how tutti-frutti climate can be. Richard Seager coauthored a paper at Lamont-Doherty Earth University. He said, "Climate variability tends to occur within patterns that span the globe, creating wet conditions somewhere and dry conditions somewhere else."[28] To give you some perspective on droughts from ancient times to present, here's a list.

Droughts by year (or close to it):

1. *The drought that helped spread humanity.*
   These droughts 135,000 and 75,000 years ago may have caused the first migrations out of Africa.[29]
2. *The drought that changed Egypt.*
   It was 4,500 years ago. Between 10,500 to 4,500 years ago, Egypt was a lush green paradise. Then the Sahara Desert moved in, which could be one of the causes of the fall of Egypt.[29, 30]
3. *The drought that possibly destroyed the Mayans.*
   This happened 1,200 years ago. Less water and less rain may have exacerbated the fall of the Mayan civilization.
4. *China's worst drought.*
   It was in 1929—1930 and has been called "the most disastrous event of the 20th century in China."[29]
5. *The drought the spread deadly disease.*
   The American Midwest, 1931 to 1939. Black Sunday or the Dust Bowl. The dust carried with it measles, influenza, and valley fever.[29]
   1941-1942 China
   1987-1989 Northern Great Plains, US
   2006-2010 Syria[31]

These are pretty much the worst droughts in our history (according to my research). All the research points to the first five as being human-changing events. At any given time, there are droughts somewhere in the world but not on that scale. The activists and their scientists predict we are going to have more and even worse droughts. US droughts are similar to those in other parts of the world. Below, I've listed the historical timeline of major droughts in the US. Keep in mind the preindustrial date of 1880.

Tenth to Fourteenth Century
According to the National Academy of Sciences, around the year 900 started a drought so bad it has been called a megadrought. Tree ring studies bear this out.[230]

Fourteenth- to Nineteenth-Century Mississippi River Valley Drought
The drought started around 1350 and lasted about five hundred years. It is said that this is what triggered the Little Ice Age and the changing climate.[230]

Nineteenth Century
    1845— Sonoma drought
    1856–1865—Civil War drought
    1870–1877
    1890–1896

Twentieth Century
    1930–1936—Dust Bowl, the Dirty Thirties, devastating drought
    1944–1951—Southwestern US longest recorded drought for that area
    1953–1957—Midwest US, Texas suffered its most severe drought in history[32]

We could go on listing droughts, but I think you can see that there were major droughts in ancient history, just as there are today. Some of them human-changing events. You can't just take the last hundred years and think you have a good picture of these events. There are droughts all over the globe. China has had devastating droughts in its long history. If you don't look at our past climate, you can't make judgments on the present or future of our planet. (Again, my opinion.)

In 1900, the world population was about 1,600,000,000. By 2019, it was 7,713,468,100.[33] Yup, the population is growing and will most likely continue to grow at an even faster pace. As I researched data and put them in graphs and charts, I realized that with increasing population and better technology, the number of reports and tracking have gotten so much better. My observation: this can form a misleading view that our storms are increasing in number. To better see any trend, we must observe our weather patterns from the point where high tech kicked in and began tracking and predicting. Collect data for another fifty to one hundred years. Oops! Then it will be too late. Right now, we don't know if the increase in storm activity is indeed due to temperature rise or simply better technology and more people around to observe the weather. But who has time to wait for the correct data results? They tell us our situation is dire. We have no more time.

Other Stuff

I'll take this opportunity to discuss some interesting facts since I already spent a long time on the weather. How do we know how much greenhouse gases humans are spewing into the atmosphere? It's a very important question and a difficult one to answer. First, what type of human activity emits all this GHG? Most *experts* have decided these are the main culprits.

Burning fossil fuels. Since the beginning of the Industrial Revolution in the mid-eighteenth century (a lot more than one hundred years ago), the demand for fossil fuels has continually increased as *population* increased. This makes CO_2e levels increase. How do we know that? Today, we know that burning fossil fuel releases CO_2e. Fact. Scientific experiments over the years have proven that, and how much. For example, we know from experiments that one ton (2,000 pounds) of coal generates 2.9 tons (5,720 pounds) of CO_2. For gasoline, there are many variables. Before the gas even reaches your car, emissions can vary from 3.4 pounds CO_2 per gallon to 6.7.[34] You are now jumping up and down with joy because you drive an electric car. I hate to be the bearer of bad news, but in most regions, your electricity is dependent on coal. Running your car on electricity doesn't reduce the CO_2 emissions very much compared to gas cars. Every time you plug in your car, you are using electricity powered by ... wait for it ... *coal plants*. Sorry. Not all areas, only those that have coal plants.

Back in the eighteenth century, all that was known was coal and wood. Coal plants were very efficient, just like they are now. Coal for warming homes and businesses. Yes, they polluted their cities and the environment. Most of it went unnoticed.

Farming and forestry. Every time we farm or cut trees down to make farmland, we are changing the land. When we change the land, we generate what is called *anthropogenic* CO_2. That is environmental pollution caused by humans. Using manure changes our wetlands, and venting landfills releases methane. Fertilizers release nitrous oxide (N_2O), another bad gas.[35] And it's not just farming. In some countries, millions of trees are cut down and burned for grazing and farming. That emits CO_2e. Look at the Amazon and Madagascar. In areas where most or all the trees have been cut down, there is a change in how the sun's heat is absorbed and reflected. Cutting down trees decreases the amount of CO_2 absorbed by the trees. Burning the trees releases even more greenhouse gases.

Even biomass plants burn wood and emit CO_2e big-time. But we're getting off track. What we need to know is how our forests are doing today compared to 1920, for example. Of course, our total acreage has gone down since 1750, but how have we been doing in the last one hundred years? We've actually been doing OK. Our total acreage in 1920 was about 735 million acres of forests. That year was America's lowest point. Since then, we have been working on this issue, and by 2016, America was at approximately 819 million acres. Even though the US and a few other countries are planting trees, the rest of the world isn't doing so well. Planting trees is now a priority for all of us. But can we plant enough for the future?

Livestock. According to UN experts(?), livestock (cows) releases 18 percent of GHG.

> This percentage includes the effects of deforestation in order to create grazing land, as well as livestock

natural methane gas emissions. These include nitrous oxide (which is 296 times worse than the global warming potential of CO_2 and methane (which is 23 times the global warming potential of CO_2.[35]

We have all heard this new idea of no more beef. We need to get rid of the cattle industry and eat plant-based artificial meat. Listen up! Humans are meat eaters. If you plan to lessen or rid the country of beef altogether, then there are thousands of other animals we will need to get rid of that also fart methane. Here is a small list: bison, deer, moose, pigs, wild boar, goats, wild mountain goats, horses, camels, reindeer, and so on it goes. Their farts aren't as strong as the cow, but add it up, and it's a lot of methane. I'm just taking progressive ideas one step further. That's all. Can you imagine what that will do to the price of real beef?

I guess that's the whole idea.

Cement manufacturing. Here's a little chemistry lesson. When you heat calcium carbonate, it makes lime and CO_2. More CO_2 is released when you burn fossil fuels to make cement. That's about it. For every 2,205 pounds of cement made, 1,982 pounds of CO_2 are released. This will become important later when we discuss renewables.[35]

Aerosols. Aerosols are particles suspended in the atmosphere. I'll bet most of you knew that. Sulfate aerosols come from burning fossil fuel. However, its effect on the atmosphere is cooling the air because it reduces sunlight. Then you have your chlorofluorocarbons (CFCs). We use CFCs in refrigeration, fire suppressants, and manufacturing.[35] It's also used in propellants, in the medicinal field, and as degreasing agents. Those with the biggest claim to fame are Freon and Teflon.[36] CFCs affect the ozone that protects us from damaging ultraviolet rays coming from the sun. Oops.

Another chemical compound is fluoroform (CHF_3). It doesn't deplete the ozone layer. It is a precursor to the manufacture of Teflon. It is also used in the semiconductor industry, fire suppressants, and in refrigeration. CHF_3, also known as HFC-23, is a potent GHG. The release of one ton of HFC-23 has the same effect as 11,700 tons of

CO_2.[37] Recently, it was found that HFC-23 emissions are higher than ever. It is 12,400 times more damaging than CO_2. So, where is it coming from? The world is working on phasing out this gas, which will leave China holding the bag.[38]

There are many compounds that are being used in many different applications, but I think these give you an idea of the scope of our human needs. What are we willing to give up?

The GND also says that our "extreme weather events ... threaten human life, healthy communities, and critical infrastructure." Every single storm I've listed so far, whether it was in ancient times, the 1900s, or today, has subjected humankind to death, loss of communities, and even loss of entire civilizations. Every big storm causes infrastructure damage, some more than others. With the increase in population, we are building more and more housing, building multimillion-dollar homes and huge high-rises. Of course, it's going to cause severe damage to property and infrastructure. That's the problem of humans, *not* the storms. To subscribe to the idea that today's storms are worse than those in the past is just not logical. The earth has had many weather events over the centuries that were far more damaging and dangerous than we see today. Here in the US, we have bad years and not so bad years. That's nature. But population continues to increase.

Right now, I think this would be a good time to talk about the Atlantic Ocean currents. No one has discussed this topic much, but I'll throw this into the mix. Do with it as you will. This won't be a long lecture! In August 2021, the Nature Climate Change suggested that research shows that Atlantic currents were indeed weakening. What happens if nothing is done?

The last time this happened was 14,500 years ago. What alters this great Atlantic current is the amount of fresh water entering the oceans from massive glaciers melting. This disrupted the ocean's currents. What happened next? Well, without the current to move warm, tropical water to the northern latitudes, the North Pole began to freeze again, and the Northern Hemisphere began to freeze. This lasted for three thousand years.

This report stated that we could see cooler temperatures as much as 14°F in certain parts of the planet. We could have permanent flooding along the coastlines; even entire cities would be flooded. Parts of Africa could see permanent droughts. This catastrophe won't happen for about 250–300 years from now with our current emissions. It is a climate disaster that won't happen overnight, and so humans face some important choices as to how we will handle this future disaster.[224]

A great man once said there are only two things to have—integrity and common sense.

3

More of the GND

This chapter continues with the text of the GND as it pertains to global warming and climate change in more specific ways, to unmask its fairy tales.

My goodness, we've gone through so much already. In the last chapter, we discussed the claim that human activity is the dominant cause of observed climate change over the last century. We talked about the predictions of extreme weather events. But the GND isn't done yet. In this chapter, we'll define *preindustrial* and what it means for temperature rise and GHG emissions. Also, what do we mean by net-zero? Let's not forget the terrible cost of climate change. We'll try to make sense of all these GND issues.

(3) Warming above 2°C from Preindustrial Levels

Here we go, folks. Our first big question is, what preindustrial temperature levels are they talking about? Is it taken from 1750 or 1880? We need that baseline temperature for any comparisons. My research found that no one really knows for sure what the 1750 preindustrial temperatures were. Scientists can only estimate. To come out and say we need to keep temperature rise to 2°C (3.6°F) below a certain point is really saying nothing. We need to know what that point is, or at least have a common definition to work from. Of course, the GND doesn't do that. That's because there is no good way to determine the 1750s temperatures except to estimate them.[39]

The IPCC did give a preindustrial date range of 1850–1900. That's a start, but by 1900, I repeat, we were well into the industrial age. Some scientists think we should go back to 1750. However, good recordkeeping on temperatures didn't start until 1880. Scientists created *climate models* for pre-1880, which is a fancy way of saying they are guessing. They made many models using slightly different data so there are many different models floating around out there. According to one research, some scientists, based on the climate models, *think* the preindustrial (1750) average surface temperature should be between 12.0°C and 15°C (54°F and 59°F). Yet, it could have been 11°C–12.5°C (52°F–55°F). The IPCC selectively chose models that only went back to 1850 (estimated temperatures) when some models went back to 1861. Wonder why? Maybe it fits their narrative better. After all, the IPCC is making a political statement, not a scientific one. Remember, many scientists use words like "I think," "should be," "best estimate," and "assume." This gives them an out in case they're wrong.

One last point. All these climate change models and estimates contain some form of error. These are temperature ranges. There is no absolute temperature average pre-1850. They are estimates, so they more than likely contain errors, which means all results contain these same errors. All this is a guess. It is their best guess. I don't blame them for any wrongdoing. It's how the IPCC used their information. I don't think we can change our entire way of life on assumptions and estimates. We need to be aware of what's going on and adapt. We definitely have more work to do.[40]

Besides all this, the Working Group I of the IPCC report used "before 1850" and 1750. The 1850 model will have one baseline temperature, and the 1750 will have a lower baseline temperature. In 1750, we were in the Little Ice Age. Since we don't have many temperatures before 1880, we rely on what scientists call proxies. Proxies are measuring tree rings, ice core, corals, and pollen to determine temperatures. But these are wrought with inaccuracies and incomplete information, and they don't always agree with one another.[41] If you take the earlier temperature, you will have a larger temperature rise because it starts from 1750. It will be a larger temperature rise than if you take the 1850–1900 temperature. But it will be full of inaccuracies. So, which preindustrial temperature do they use? The 1750 temperature would give the powers that be a larger temperature rise and give the impression of the grave nature of the situation we are in. But then the earlier temperatures can be challenged.

Here are the *average* temperatures scientists calculated for three significant time periods:

1750 (best guess)	13.4°C (56°F)[42]
1880–1900	13.7°C (57°F)[43]
2017	14.7°C (58.5°F)[44]

These temperatures are not absolute, so we can't pinpoint an exact temperature rise. It could be 0.6°F to 1.2°F. Again, I'm only giving an example. Who knows what the real temperature rise is. That's why I question some of the scientific data they are attributing to this warming trend.

(A) Mass Migration

If we follow the premise that the world is going to heat up and our glaciers will completely melt, to the point that coastal cities will be under water, then the oceans will rise and take over the land, and we will have nowhere to migrate. Let's not go down that road just yet, however. If it's the case that oceans rise a lot, then humans will have to migrate. If there are multiyear droughts in the farm belt, we may have to help farmers and ranchers migrate to better pastures, like in the 1930s during the dust bowl. People will probably have to migrate to areas that are better suited for farming, ranching, and so on.

This would not be the first time humans had to migrate. Except now money is involved. The first humans migrated and spread all over. Around twelve thousand years ago, about the time of the last great ice age, people from Siberia migrated across the land bridge to North America. Get my drift? Do *we* really want to migrate? Probably not. Humans are now financially trapped in a desperate need to save their fortunes from disappearing into the ocean realm of Neptune. I suppose humans think they can keep the status quo forever. I guess our government thinks it can get *all* the countries of the world to come together like one big happy family and help one another out. I doubt that will happen anytime soon. That's only if you assume temperatures will continue to rise and rise as all the glaciers melt away and the oceans take over the land. Doesn't that just make your hair frizz?

Part of the problem is that the human population is expected to *double* in the next sixty-three years. Then we will be in big trouble. We (the world) won't be able to produce enough fresh produce, meat, or fish to feed the world's population or produce enough energy. Again, I don't know for sure, but it only seems logical. Unless the earth has another extinction event, the GNDers believe the temperature will simply continue to rise. Oh my!

There is another climate scenario. Actually, there are many climate scenarios. The shutdown of the global conveyer belt of the oceans will bring on a mini ice age. A lot of the climate change advocates don't believe this will happen. It doesn't fit into their agenda. Their reasoning is that even if the conveyor belt slows or stops, our temperatures will be so high that it won't trigger an ice age, or maybe one so small it won't be noticed. That's all the screamers can say to justify their agenda and push their renewables.

The question is, can temperature rise affect the ocean's circulation?[45] It's called the Thermohaline Circulation and the Great Ocean Conveyor. This global circulation is very important to the climate in different regions of the planet. How does it do that? All the currents are wind driven as they move around our planet. Let's start in the Atlantic Ocean. The cold ocean water that is on the surface gets saltier as the ocean evaporates, and salt is removed when sea ice forms. This colder, saltier water is denser, so that makes it heavier. It drops deep into the ocean and travels along the bottom until it reaches a place where it can surface. That's usually near the equator, around the Pacific and Indian Oceans. Heat from the sun now warms that cold water that has surfaced, and the water proceeds to evaporate, making the water saltier again. Water then moves north and joins the Gulf Stream, another strong ocean current. This warm water moves up the US East Coast. It crosses the north Atlantic, releases heat, and warms Western Europe. The ocean current releases its heat, gets very cold and dense again, and sinks ... again. And the cycle goes on.[45] Right now, there are scientists saying the conveyor is slowing down, and that could trigger a very cold spell for the areas north of the equator.

In comes temperature rise. It causes the glaciers to melt, so more and more fresh water enters this conveyor system, cools the water, and lowers the salinity (saltiness). This and other conditions have slowed the global conveyor system by 15 to 20 percent in the twentieth century. An area in the North Atlantic was the coldest on record since 1880. Scientists think this could be due to the weakening of the conveyor system. I've heard that the Atlantic system could flip within the next few decades—faster than the scientists ever thought.

Not all scientists agree that the circulation systems are weakening, but if they are, I'm sure they will say it's human caused. That's some kind of silly. This is where scientists disagree. Some say that even if the conveyor shuts down altogether, we won't have an ice age because our temperatures will simply be too high. Other scientists aren't so sure about that. The shutdown of the conveyor system could likely bring on colder temperatures and then a mini ice age.[46] No one knows for sure. I like to think of it as when one side is throwin' their weight around, they should be ready to have it thrown back by somebody else as well.

Trying to predict what will happen is nearly impossible. The Great Ocean Conveyor system and the Gulf Stream are very complex systems. Scientists don't all agree on it and all the variables that make up these systems or how they work. There is still a lot of debate over it. One thing we can be pretty sure of is if the circulation system does collapse in the next forty or fifty years or one hundred years, we will have big troubles even without an ice age. Colder temperatures will most likely influence the food chain and fish distribution. According to one scientist, "It is a complicated system and we can't make any predictions." The bottom line is that people are reluctant to migrate because of economy … money. Money seems to get in the way of a lot of stuff.

(B) More Than $500 Trillion in Lost Annual Economic Output in the US by 2100

All I can say to that is balderdash. Someone has to make a lot of assumptions to come up with that *fact*. And they are stating that as a fact. The truth is nobody knows what state we will be in by 2100. That's eighty years from now. Who knows? I've read how scientists are working on locking up carbon. People are working on decarbonizing cement and certain plastics. We need more of that kind of research because most countries realize they do need fossil fuels.

To scare the people with such unsubstantiated statements and present them as fact is hellacious. We have had losses due to storms, hurricanes, and droughts since humans stood on two feet. We always have. We always will. It's outrageous to say our economic losses are due to climate change. That is an excuse for poor management. It's simply the mega rich and political elites who are desperate to save their investments. Do I care about the rich losing a little? Not really.

I do care about the hardworking who may need to migrate from the coastal areas, and farmers and ranchers who may need to migrate to better areas. Otherwise, to me, it's simply a scare tactic of the elite and governments who need to terrify everyone because they themselves are terrified of losing power.

(C) Wildfires Will Burn Twice as Much Forest by 2050

Here is the IPCC theory, and it is a theory:

> Wildfires that by, 2050, will annually burn at least twice as much forest area in the western United States than was burned by wildfires in the years preceding 2019.

Let's break it down. First, 2050 is three decades from now. One article assures us wildfires will be six times worse.[47] That's pretty specific. How did they calculate the future of fires? I'm not sure, but I think there are some fancy formulas and extrapolations that went into this, or maybe just guessing. We're going to take some data and a few forestry models and see if we can make any sense out of this. I presume they are using the acres burned in the west, graphing it and extrapolating it into the future. Let's see if that works. Or maybe they are using some magical mathematical formulas.

First, we need to know how many acres were burned in the west every year. We'll start at 1937. The data prior to 1983 is not very reliable.[48] That's the vertical line on the graph below. I couldn't find the acreage burned per year, but I did find that in 2018, the western states were responsible for approximately 80 percent of the US's total acreage burned.[48] How did I figure that? Data shows for 2018, 1.7 million acres burned in the east, and about 7.0 million acres burned in the west. Now, we can calculate the percentage. I'm going to do a little of what the IPCC does. A little creative calculation and assume every year will be around 80 percent, give or take. It may go up and down every year, but I will take 80 percent as an average. So, I'm going to graph four lines: the number of acres burned in the US, acres burned in the west, acres caused by human carelessness, and acres burned caused by nature. Most of the data says 80–90 percent of wildfires are caused by human carelessness. Then I will try to extrapolate that data and see if we come up with any future prediction.

The graph is acres burned over time. Time being 1937 through 2019.

Fig 6 [49]
Data points for Fig 7 - Acres burned – Millions (mil)

Year	Total burned In the U.S. 1.	80% Burned in west 2.	90% Human caused 3.	10% Nature Caused 4.
2019	4.6 mil	3.7 mil	3.3 mil	0.4 mil
2018	8.7 mil	7 mil	6.3 mil	0.7 mil
2017	10 mil	8 mil	7.2 mil	0.8 mil
2016	5.5	4.4	3.96	0.44
2015	10	8	7.2	0.8
2011-14	6.5	5.2	4.7	0.5
2006-10	6.8	5.4	4.9	0.5
2001-05	6.3	5	4.6	0.4
1996-2000	4.7	3.7	3.4	0.3
1991-95	2.6	2.0	1.8	0.2
1986-90	3.3	2.6	2.4	0.2
1981-85	2.4	1.9	1.7	0.2
1976-80	4.1	3.3	2.9	0.4
1971-75	2.7	2.2	1.9	0.3
1966-70	4.7	3.8	3.4	0.4
1961-65	4.2	3.4	3.0	0.4
1956-60	4.4	3.5	3.2	0.3
1951-55	10	8	7	1
1946-50	18	15	13	2
1941-45	26	21	19	2
1937-40	30	24	22	2

Data points for figure 7

Fig 7 Acres burned vs Time

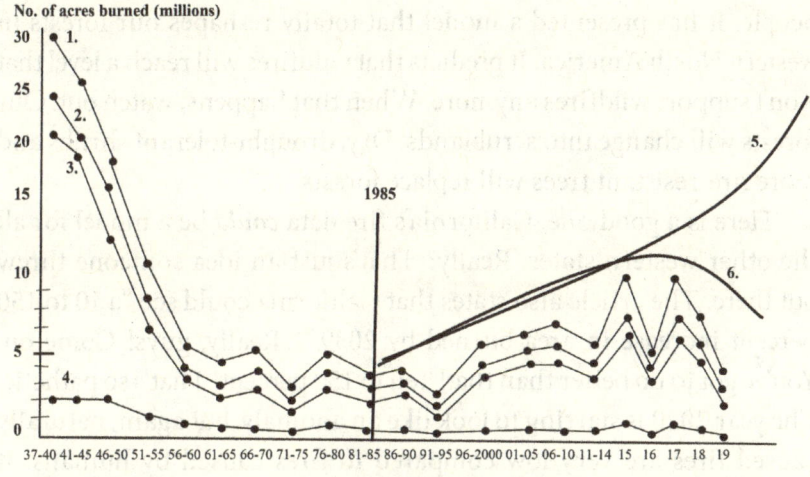

Acres burned over time

Legend:
1. Total acres burned in the U.S.
2. 80 percent burned in the western U.S.
3. 90 percent human caused
4. 10 percent nature caused
5. Wildfire projection model from climate change proponents
6. Wildfires as it actually is as of 2019

The point of this graph is that wildfires caused by nature remain steady and quite small. These are the one caused by *climate change*. The only conclusion I can make is that population increases and encroachment into the forests increase the number of fires by careless people. And that's not climate change caused by fossil fuels. If you take 1985 through 2019, it appears to go up, but after 2015, it appears to have several large drop-offs. In 2019, it also dropped. This may be a little confusing because there are so many predictions out there. However, if you look at the graph for 2015 through 2019, it closely matches a graph released by the National Centers for Environmental Information (NCEI) for the same years.[50] One article predicts wildfires in at least eleven states will increase in the annual acres burned by 500 percent.[51] This article is a real doomsday predictor meant to scare people. It has presented a model that totally reshapes our forests in western North America. It predicts that wildfires will reach a level that won't support wildfires anymore. When that happens, watch out. Our forests will change into scrublands. Dry, drought-tolerant shrubs and more fire-resistant trees will replace forests.

Here is a good one. California's fire data *could* be a model for all the other western states. Really? That's just an idea someone threw out there. The article also states that California could see "a 10 to 150 percent increase in area burned by 2039."[51] Really, guys! Come on. You've got to do better than that! Ten to 150 percent! That's so pathetic. The year 2020 is starting to look like an anomaly, but again, naturally caused fires are very low compared to fires caused by humans. It makes me think that global leaders want to scare us with their fancy magic to gain support for renewables and their GND. Look at the fires in Northern California. Climate change! Climate change!

Out of the 8.7 million acres burned in the US in 2018, about seven million acres burned in the west. What is the west? According to the Congressional Research Service (CRS) and the National Interagency Fire Center (NIFC), Alaska, Arizona, California, Colorado, Idaho, Hawaii, Montana, Nevada, Oregon, Utah, Washington, and Wyoming are considered the western states of the US.[48] The GND didn't define the west. I guess it doesn't matter to them. It should to us.

I'm just quickly going to throw this into the mix. When people crossed the land bridge from Siberia, they brought fire with them. They didn't have the tools to cut down trees, so they set fire to the bottom of them. They used fire to encircle large animals to make it easier to kill them. They used fire to clear large areas of land to make travel easier. Fires were used in warfare. The Native Americans used fire to reshape the ecosystem. Whether we like it or not, humans were, and continue to be, a factor in reshaping the landscape around the world.[51]

The next thing is "the years preceding 2019." What does that even mean? How many years preceding 2019? Do we go back twenty, fifty, or one hundred years? They didn't give us a starting point. If there is no starting point or date range, they can't predict what's going to happen. I did a bunch of research on wildfire statistics, and it took a lot to figure out statistics for just the western states. As much information as I have given here, the only conclusion is that we can't predict the future of wildfires. If humans continue to be careless with the environment, then wildfires should increase as the population increases. That, my friend, has little to do with fossil fuels unless you count the few accidental fires caused by sparks from chainsaws, vehicle backfires, and downed power lines. To save our lands, we must teach people and our children how to care for the land and the ecosystems that go with it. We must stand up together against those who would pull the wool over our eyes. Look at that little black line at the bottom of the graph. That's the number of climate-caused fires. Even with all the screaming over temperature rise and climate change, those numbers seem to hold steady and low. Just get these humans out of our forest ecosystems! Maybe humans who wish to enter the divine peace of our forests should be required to take classes and a knowledge test. And stop building into the forests!

(D) Ninety-Nine Percent Loss of All Coral Reefs on Earth[52]

There is no doubt coral reefs are in trouble worldwide. It's not just the reefs; it's the complex ecosystems that go along with the reefs. Nothing else rivals coral reefs as the most diverse ecosystem on earth. They have extremely specific needs. To survive, they need clear, clean, warm water. But that's not the total picture. The last fifty years have been ridiculously hard on our reef systems. Not just from temperature rise but from dynamite fishing, dumping waste and trash into the oceans, and dropping and dragging anchors across coral reefs. You can find pictures and videos from around the world of people dumping their trucks of trash into the local rivers. Large ships, like cruise liners, dump their garbage into the oceans, not to mention tankers, container ships, and barges. You can find pictures of coastlines with miles of trash filling the beaches. I did read an article about how cruise ship companies are working on this issue. We'll see.

Marine regulations are so crazy because every country has its own ships and decides how they will manage their waste. The differences in jurisdiction as well as the definitions of waste and waste treatment make enforcement nearly impossible. I have to quote this next piece of information in its entirety because you all need to read it and see what your take is on it:

> Cruise ship discharges of solid waste are governed by two laws. Title I of the <u>Marine Protection, Research and Sanctuaries Act</u> makes it illegal to transport garbage from the United States for the purpose of dumping it into ocean waters without a permit or to dump material from outside the U.S. into U.S. waters. Beyond U.S. waters, no MPRSA permit is required for a cruise ship to discharge solid waste. The routine discharge of effluent incidental to the propulsion of vessels is explicitly exempt from the definition of dumping in the MPRDA.28. The <u>Act to Prevent Pollution from Ships</u> prohibits the discharge of all garbage within 3 nautical miles (5.6 km) of shore, certain types of garbage within 12 nautical miles (22 km) offshore, and plastic anywhere. It applies to all vessels operating in U.S. navigable waters and the <u>Exclusive Economic Zone</u> (EEZ).[54]

Does that sound like protection of the planet's oceans? And the US is leading the way. Other countries have even fewer restrictions, which is why we see many beaches filled with god-awful human garbage. Just imagine all the ships that sail the seas from countries all over the world. How many tons of garbage are dumped into our oceans? I don't think they want us to know all this. We know, but many people don't register it in their brain.

The reef foundations get torn apart from dynamite fishing and dragging anchors. This is all human-caused destruction. But is it due to fossil fuels and temperature rise? It's due to humans not caring enough to keep their reefs and oceans clean and healthy. A part is due to ocean levels rising.

We need to get all humans, or at least 99 percent, to stop polluting and destroying our oceans. One article reports that the latest numbers show that "27 percent of monitored reefs formations have been lost and as much as 32 percent are at risk of being lost within the next 32 years."[53]

Another article says 50 percent of reefs have been lost in the last thirty years, and 90 percent may be lost in the next century. This article goes into economic losses from fishing and tourism but blames it solely on climate change. Temperature rise has had an effect, of course, and we need to address that. There's very little mention of these other causes of reef loss. The governments never bring up these issues. There's no money in it. I get so infuriated when these activist politicians, who are presumed to be stewards of the environment, ignore half the story. They just want us to believe that global warming is the only evil dragon. Buy renewables!

Even if we get rid of fossil fuels, it will probably take hundreds of years, maybe even thousands, to reverse the damage to coral reefs. The GNDers don't care about stopping reef destruction from dynamite fishing, trash dumping, overfishing, and general pollution of our oceans. I do have to credit those few nonprofit organizations that work on those issues. Get with it! We need the truth, the whole truth about what's happening around our planet. Yes, we need to work on global warming in terms of carbon emissions, but that won't help our coral reefs if we don't tackle these other issues. It's like putting a Band-Aid on a gunshot wound.

(E) More Than 350,000,000 People to Be Exposed Globally to Deadly Heat Stress by 2050

Humans emerged from Africa about forty thousand years ago. They migrated during a very unstable climate. As they moved around from place to place, they developed new traits. For instance, as humans migrated north toward the Arctic, some of them became fair-skinned and developed lighter hair. One of the researchers into early humans, Richard Potts, says that "human adaptability came from environmental change over the long term."[55] Which environmental change pushes migration the most? Rain and floods, heat, snow, or ice?

It seems that heat is the big driver for human migration.[56] The body works to maintain homeostasis. Everything stays even. For instance, thermoregulation is one type of homeostasis. The body tries to maintain a temperature of 98.6°F. If it goes too low or too high, the body can shut down if there is no intervention. Humans have adapted to a variety of living conditions in which hypothermia (low body temperature) and hyperthermia (high body temperature) are common but have adapted using clothing, shelter, and fire. They migrated to areas that promoted their well-being, perhaps near rivers, lakes, or the coast.

There are many diverse cultures on our planet. They migrated and eventually adapted. For example, the Inuit have more blood flowing into their limbs than people in warm climates. Research even shows that people in cold climates have shorter limbs than those who live in warmer climates.[55]

Humans have been adapting and moving since the dawn of man. It wasn't a huge problem back then. Even in the last three hundred years, migration was working. The Irish migrated to America during the potato famine. Native Americans migrated with the seasons to follow the herds. Our modern complication to all this (especially the more industrialized countries) comes in the form of money. What the world has now is an economic system that doesn't mesh well with the natural migration systems because of land ownership. Farmers, ranchers, and fisheries can't migrate anymore to find richer pastures.

The upshot is that we need to go beyond the crazy rhetoric of the politicians and develop an understanding of the impact to our society of sea level rise, desertification, and drought. We need to really try to understand, through research and education, how all this affects our fishing, agriculture and ranching and how we (as a people) can help one another through whatever comes our way.[56] From my own reading, I found that all these natural disasters initiate help from government agencies, but heat stress rarely attracts attention. In 2010, high temperatures in Pakistan led to a 13 percent decline in their wheat yields. They got no help in terms of money from their government or international relief.

The last part of the research says that people who did not own land were more likely to move. So, in the modern age of great technology and science, it's still money that rules the roost.[56] Again, it's big money politics wanting to scare everyone into pleading for the GND.

(F) A Risk of Damage to $1 trillion of Public Infrastructure and Coastal Real Estate in the US

Notice the title says, "A *risk* of damage ..." They are not telling us there will be damage but that there might be damage from rising ocean waters to the tune of $1 trillion. Maybe there will be damage. I'm not giving my tax dollar to a maybe. Coastal property is very, very expensive, and roadways can be hard to build. So, they make up a really big number. They don't know how much financial loss or damage there might be because they don't know what the coastal future is going to be. That's my opinion.

Another opinion of mine is that the GNDers try to rattle the rich, like a rattle snake on a rat, so it scares them enough to get on board and donate, donate, donate and support mega-high taxes for the rest of us to help infrastructure. Maybe they do. I only hear about them building very expensive, ecofriendly cars and spaceships. Every year, we have infrastructure damage somewhere along our coast. It's been happening since Europeans arrived and even before. What are you going to do when Los Angeles meets up with San Francisco? It seems like it's going to happen. Can you imagine all that infrastructure damage and loss of real estate along the way? I'm not complaining about the really rich, I think they do a lot for the environment, but then they also do some stupid things.

The last thing in this section I want to make clear is that forsaking the fossil fuel industry is not going to happen any time soon unless we want to return to the Stone Age. Well, it may not be quite that bad. I will refer to chapter 8 of the IPCC Special Report on Carbon Capture and Storage (CCS). It kind of makes my point on all this scientific discussion on how to handle climate change.

To save us all this money in damage, we need to lower the carbon in the atmosphere so it lowers the temperature and stops glacier melt and water levels will supposedly go back down. That's the bottom line. It may not be feasible in the next few years. And this is why. It asks two questions: "How much will it cost?" and "How do Carbon Capture and Storage technologies fit into a portfolio of greenhouse gas mitigation options?" (But I love carbon capture.) All they are doing is yapping.

This IPCC special report goes on and on with fancy graphs and data on just this one idea for lowering carbon in the atmosphere. Several statements made a huge impression on me. They present three risk factors in the carbon capture plan:

- Ecological risk. Leaking from the storage
- Financial risk. Carbon prices might exceed expectations.
- Political. Institutions might manipulate the regulatory systems in their favor. (IPCC)

Wow! Who didn't see that coming?

I still like carbon capture. Get the politics out of it. Their answer to this is that we need to closely follow future scientific debate on these issues. OK. More discussions while time is flying by. They are not done discussing whether carbon capture is a good idea or not. The last thing in this report is a doozy. Because energy prices and climate change policies are so fluid, there isn't current literature available. That means their climate models, data, and graphs kind of make this about as useful as letting your hedgehog do the thinking. And finally, to the last quote, "This suggests a need for a continuous effort to update analyses and perhaps draft a range of scenarios with a wider range of assumptions ..."[57] There you have it. Assumptions!

(4) Must Keep Global Temperature below 1.5°C above Preindustrial Levels

Earlier, I discussed the problem of defining what preindustrial temperatures were. The problem is we don't exactly know what those temperatures were. Therefore, it is nearly impossible to talk about a 2°C or 1.5°C rise if you don't know where you are starting from. Some reports have tried to convince us we don't need to know the preindustrial temperature. Well, that's like if you stop thinking and then forget to start again. Enough with that. Let's trek on. It's easier to think of preindustrial time as post-1880, like many scientists are now doing. It's not the true picture, but at least it's something to work from.

(A) Global Reductions in GHG of 40–60 Percent from 2010 Levels by 2030

Well, butter my butt and call me a biscuit. Are they flat-out looney? It's not that it shouldn't happen. It would be great if the entire planet suddenly got together, with a huge kumbaya, to get it done, including China and India. Even if the US reduced its emissions by 60 or 70 percent, it won't do much of anything to lower *global* temperatures or GHG worldwide. And don't bring up that America has to lead the way. We have led the way, and by our innovations in the last hundred years until recently, when a certain Democratic presidency decided we were not an exceptional nation. They call all our achievements racist, homophobic, and so on. We have been reducing our emissions, people! Besides, we would have to come up with innovative ways to either wean off fossil fuels by tomorrow or find a way to neutralize the carbon we are emitting. We could do the second one if not for the elites who want to continue their money train. We have ideas. We have prototypes. Canada even has a good prototype. We need to get them built and in use, thank you very much. I'm not talking about the renewables we have other technologies to help reduce carbon in the atmosphere. Some we will discuss later in this book.

(B) Net-Zero Global Emissions by 2050

Don't get me wrong. I believe in lowering our carbon emissions. The GNDers haven't worked out a good plan that can be set in motion that works! They want renewables like wind and solar. Renewables won't work in the long term. I'm going to try to tell you why. I did my homework. If we are to fight this rise in temperature, we must start putting our research into better and more creative alternatives because the world doesn't have really good ones. According to the IPCC, these are the top recommended renewable energy (RE) sources:

- wind
- direct solar
- geothermal
- hydropower
- ocean energy
- bioenergy (really bad)

Not on their list is nuclear energy. I guess that's not an option. We'll go into more detail in chapters 5 and 6 on these renewables because that's pretty much what the GND advocates. The US has some geothermal and hydropower. Then the last two to bring up the rear, ocean energy and bioenergy. We'll see how these all fit into the puzzle of renewable energy. This is one of the most important sections because we all need to understand the carbon footprint of all these renewables and their life span.

Getting back to net-zero by 2050, this last part is based on an article by Roger Pielke, contributing editor for *Forbes* at the time, and data from BP *Statistical Review of World Energy*.[152]

The DNC platform for the 2020 presidential election promised a net-zero carbon emissions target by 2050. I'm just stating a fact here. So, let's throw around some numbers. I used this article, and I'll summarize it for you. We need to look at a few things when discussing net-zero by 2050. What we need to look at is the rate that *carbon-free energy* is created versus the closure of the fossil fuel business and energy.

First, look at the global fossil fuel consumption. In 2018, the world consumed 11,743 mtoe (million tons of oil equivalent). That's coal, gas, and petroleum. They spewed out 33.7 billion tonnes of CO_2. Keep following me. For those emissions to get to net-zero, we would have needed to replace around 12,000 mtoe of energy expected for 2019. And that day has clearly passed. Gone bye-bye. I only mention this because my reference article was written in September 2019. We're far behind. We're kind of like a donkey trying to catch up to a racehorse.

I've given you the first set of numbers. Here's the second set of figures. From the time the article was written, there were about 11,051 days to January 1, 2050. This number is very important because the number of days combined with the first set of numbers will give you a big fat surprise, like finding a skunk in your bedroll.

To globally reach net-zero CO_2 emissions by 2050 would require the use of more than 1 mtoe of carbon-free energy per day, which is around 12,000 mtoe for 11,051 days. That starts the day after the article was written ... September 30, 2019. Besides all that, getting to net-zero also has to correspond to getting rid of 1 mtoe of fossil fuels per day. Those numbers don't stay stable since we have to include an ongoing increase in demand for energy every year.

Some people probably have a hard time understanding what mtoe stands for. Let me try to give you a visual. Net-zero by 2050 would mean the need for building 1,500 2.5 MW wind turbines over three hundred square miles *per day*, starting September 31, 2019, and continuing until 2050. Remember, these are global numbers. According to BP (British Petroleum), in 2018, the US used 1,900 mtoe of fossil fuel energy. Doing the math ... the US would need to build one nuclear power plant about every six days from September 31, 2019, to 2050, and that's not accounting for the increase in demand.[152] For those who hate nuclear, let me convert that to wind turbines.

Fig 8 Nuclear vs Wind Power

1 Nuclear plant – 1,154 MW	=	2,000 2.5MW wind turbines
9,100,000 MWh/year		9,100,000 MWh/year
90% Capacity Factor, Offline only for refueling and maintenance.		25% Cap Factor, variable due to weather
Approx. 50 acres with one sq. mile buffer zone. [153]		Approx. 3,200 acres

Nuclear versus wind power

That's it in a nutshell, folks. We will destroy millions and millions of acres of land, forests, and ocean bottom for our current renewable energy needs. Why don't we ever see the numbers? I think it's pretty obvious. How could the GNDers ever truthfully answer any of our questions about these numbers? They don't want us to know. They don't want to know. They run only on ideology.

This pretty much ends the GND's take on climate change and what will happen if we don't listen to their apocalyptic warnings. But the architects of the GND still aren't finished. They want to radically change society as we know it. Brace yourselves. The next chapter is about our economic transformation and their sense of justice.

An angel appears and offers you wisdom, beauty, or $10 million. Choose *one*.

4

Whereas and the Longest Sentence in the World

This chapter takes us to the end of the GND and is riddled with "whereas" this and "whereas" that, so it makes it hard to separate theories from facts, but that's what they want. So far, we have fought wildfires, looked at temperatures, and investigated what a net-zero policy would do to our economy. And the GND still isn't finished. It tries to sell us on the idea that transportation, health care, education, and jobs are all related to climate change.

There sure is an army of *whereas*'s in this chapter. By that, I mean the GND uses this word to create supposed facts and then tell us what we have to do to resolve everything. It uses *whereas* no fewer than seven times in order to direct us to lower our emissions through "economic transformation." In this chapter, we will only talk about issues as they pertain to their climate agenda and not how they try to link social issues to climate. We know that's stupid.

There are a lot of social issues I will not be covering because I personally don't believe they have anything to do with climate change or temperature rise. That's just an excuse for horrible management. Besides, the use of *whereas* assumes what they are saying is true and we should swallow it. Are the authors of the GND trying to tell us that all our social issues of education, health care, and housing are all due to climate change? Right now, they are saying all of our social issues are due to racism, then Putin and now the Republicans. Horse pucky. Humans have had these issues since before the Dark Ages. We shall not veer from our path by discussing the social issues of our time that the Left created. We'll discuss these issues only as they pertain to temperature rise and climate change.

I was going to give an example of a "whereas" from the resolution but decided against it. It was getting too complicated. I think it's wiser if we just stick to climate change, temperature rise, and renewables. Have you read the entire fourteen pages of the GND yet?

The United States is a big country with a big population and is getting bigger every day. The larger our population gets, the more demand for power and recourses there will be. My plan here is only to discuss the GND in terms of temperature, climate, and even population. If it doesn't seem worth the effort, then it probably isn't.

There are a few things worth mentioning in this section. The first pertains to lowering emissions through *economic transformation*. What the heck is that? It's a process of moving jobs (workers) from one job to another. By definition, it's *supposed* to be from a lower productive job to a higher productive job. Then raising those higher productive jobs even higher. Like moving coal and oil jobs (high wage) to manufacturing and energy jobs (lower wage).[154] Oops! Well, that sounds great, but that's kind of like government forcing or bribing people out of their jobs and dumping them anywhere government wants. Sort of like what China does. However, I'm not sure the employees from the Keystone XL Pipeline got any of that economic transformation.

What does all this moving around have to do with lowering emissions? There's only one thing I can see. Remember when Hilary Clinton said something to the effect that she would end coal mining and retrain all those miners to do other jobs? I figure that's economic transformation. I'm not bashing Hilary; there was a lot of great-sounding stuff in her plan. It's just nothing happened. They were going to force perfectly happy and well-paid miners (we need coal for every single renewable project) to leave their jobs to learn other jobs so we could get rid of fossil fuels. But what jobs? High-tech computers or manufacturing? Renewables? How about window washers? Jobs they don't care about and have no interest in. Coal miners are proud of their jobs. And how long would retraining take anyway?

So, what happened? I don't think coal miners got much if any retraining. What's going to happen when contractors discover they can't build any renewables because they don't have enough steel or cement? I guess it doesn't matter when you're trying to convince people they're saving the planet. (Ahem, cough, cough!) I think you see where I'm going with this. Get all this mining and steel manufacturing out of the US so we look good with our emissions. But then import them. Let some other country do the mining and manufacturing so we can say how great we are with lowering our emissions. Observation. Just ponder that for a while. If you want a good drink of water from the river, always drink upstream from the herd. Let's leave it at that.

Here comes a huge *whereas*. We have several related crises here.

(1) What a laundry list of things that are apparently all related to climate change. Me thinks that is the point of this section. The GND states *life expectancy* is getting worse, and meanwhile things like housing, education, transportation are not available to everyone. The killer is that, according to the GND, all this stuff is "inaccessible to a significant portion of the United States." Stop just a second. The GND is presenting all this information as *fact*. It's not all fact. This is only put forward as fact. The fact also says that all these are supposed to be due to climate change. I'd like to see their data and documentation. There are, I admit, a lot of homeless people. There is housing for many of them. A lot prefer the street. But they have access to healthcare if they want it. Cities have a lot of programs for the homeless if they want it. The question is: Why do we have homelessness? That's a complicated question but it certainly not due to climate change.

Allow me to tackle two issues here. *Life expectancy* and *significant portion*. According to health insurers, there are a bunch of numbers being thrown about regarding life expectancy, like blowing bubbles in the wind. They bandy about all sorts of numbers, but I finally found their predictions. For men, the US life expectancy is projected to increase from 76.1 to 79.5 years by 2030. For women, it is 81.1 to 83.3 by 2030.[155] Then by 2040, the projection is a decrease. I don't know why. Finally, the report says all this is not written in stone. The future is unwritten. None of this has anything to do with climate change. Of course, the GND is trying to imply that our life expectancy is directly related to climate change. However, emissions have been going down in the US.[156] If emissions are decreasing overall, then shouldn't our life expectancy be getting better? Life expectancy also depends on our diet, exercise, smoking, drug use, drinking habits and seeing a doctor regularly.

The second issue has to do with *significant portion*. It just means most of the population doesn't have access to a bunch of stuff. These are solely social issues, and I don't believe these have anything to do with climate change. It should be dealt with by the communities, the states, and the federal government. Not by this book.

(2) Just about this entire section is social issues. It talks about a four-decade trend. There will be no discussion of the social issues in sections (A), (B), and (C) here. The only mention of climate change is in section (D): "inadequate resources for public sector workers to confront the challenges of climate change at the local, State, and Federal levels." That doesn't explain anything, so I'm going to put my two cents in.

A public sector worker is someone like a teacher, fireman, policeman, or transit operator. What a bunch of fish farts. There is nothing to stop people from challenging those in power to do something about climate change if they are so inclined. We can call our state and federal legislators and let them know what's on our minds. They may or may not pay attention, but we keep on trying. We can go to town hall meetings. We can form groups and plan demonstrations in hopes they listen to us. Teachers' unions do that all the time and usually get what they want. Lastly, *we can vote*.

The only thing that stops people are people themselves. Go to the city council meetings. This idea the Left is putting forth is to tell us we can't; therefore we shouldn't even try. In 2019 and 2020, there's been a lot of confronting going on. Not all good. And with no constructive ideas. And I don't mean burning down the local Burger King. Let's start getting some constructive ideas swirling around our government agencies and stop being treated like some kind of imbeciles. Because we are not. We are *the people*.

The only question I have is, what resources do they need? The weather gets hot, then cold. It's dry, and then it rains. Then we get snowstorms, tornadoes, and more. These weather events are not climate change. It's just another one of those nonsensical statements that doesn't mean very much but sounds really good.

All the rest of the section (3, A, B, C) are social issues. We the people vote for Congress and need to vote for people who will actually help solve our issues.

Another whereas. This paragraph is one very long sentence. We have "climate change, pollution, and environmental destruction" I'm going to stop here for a few minutes and talk about these issues because the rest of the paragraph is filled with social issues. See how the GND intertwines climate change with social issues to make it look like our terrible social issues are the result of climate change? According to the GND, climate change, pollution, and environmental destruction are all caused by humans. Every race, gender, and age group has caused pollution and environmental destruction, although environmental destruction is also caused by weather events and the building of renewables. I'm still out on climate change. They are not using the term *climate change* in its correct sense.

The GND is saying that climate change exacerbates all our social problems. I don't think so. If you look back through the ages, humanity has had nothing but social problems. We have not created anything new. Maybe that's one reason the GNDers only want to look back one hundred years. As humans, we have not learned to work together as communities, states, political groups, or countries. You can't blame climate change or pollution or environmental destruction for our ongoing human hatred for one another. And it doesn't look like it's going to change anytime soon. You can't blame climate change for problems we ourselves create through our policies.

Whereas! Whereas! Whereas! It's driving me crazy! Let me get my brain together. We all have pieces of crazy in us; some of us have bigger pieces than others. And some of us have ginormous pieces.

(1) According to the GND, climate change constitutes a direct threat to the national security of the United States. I managed to find a report from May 2015 on national security and climate change.[157] In 2015, President Obama's White House put out this report. It's interesting as to why they did. How about to scare the bejesus out of us and make us conform to one way of thinking? They must think we are a bunch of lemmings. We are not. We are the people with the freedom to think for ourselves. They are the lemmings. I believe that if one of the GNDers jumps off a cliff, they all will.

Climate change is predicted to strain economies and societies around the world. That's a *prediction*. It's not written in stone. We have a 1.0°F to 1.33°F rise in temperature, and we're in 2020. According to the IPCC, we need to stay under 2.7°F from preindustrial temperatures. And there are other temperatures roaming around. I found a report written by Iberdrola, an environmental group that is dedicated to saving the environment.

The first thing I took away from this paper is that they emphasize climate change as already having happened. I don't happen to agree with that. But that's my opinion. Remember, climate change is the *long-term* change in our weather patterns that don't go back to a normal state. We don't have that. Having said that, if there is a continuous rise in global temperature, it will eventually affect our planet's climate.

If temperatures continue to rise, whatever the cause, it will eventually become a threat to the economic stability of the planet, affecting every country. But to call it climate change, I think we need to see heatwaves in the winter and in places that never had them. Hurricanes and typhoons during the wrong time of year or all year long. Major droughts that continue to last for many, many, years without lifting.

We need to see permanent changes in our weather patterns to call it climate change. What we are seeing now could be the precursor of a future climate change. I'm not saying to ignore what's going on—just to know exactly where we are in the equation. The other problem in all this is the expansion of population. When we said eleven billion globally by 2050, it means more millions will end up in poverty. That includes the US. Increasing population has nothing to do with climate change. Here are a couple of things to end with. The world should put the environment front and center. It's one of the things that could save us. We need to be honest as to what things are hurting our environment.

The last thing this report suggested was to apply "prices," although I say taxes, "that will act as a deterrent to using fossil fuels." Their point is by having low prices, it doesn't encourage research into new technologies that would lower emissions.[223] My solution would be to tax only the mining, manufacturing, and companies that emit GHGs, not the people who need natural gas or gas for their cars, and not allow these GHG emitters to pass the cost to consumers. Let's get solutions and put them in place before we kill fossil fuels. They're putting the cart before the horse. And politicians will never fail their donors.

The GND goes on to say that "climate change will change the nature of the U.S. military missions." Are they talking about the military shifting to a more humanitarian effort? We're already doing that. I know more military has been deployed to the Arctic. Why? Is it because Russia is doing the same? Maybe. We know the Biden administration is bringing back our troops from Afghanistan. What a disaster that was. Does the Left have other plans for them? I guess we'll see. The GND doesn't explain anything.

This next statement gets me. "Demand more resources in the

Artic." Why? And what resources? Scientists? The military? Global scientists already monitor glaciers, water temperature, salinity, and more. What does the government want? What are they looking for? The GND doesn't say. Do they want more scientists? And what does a military presence have to do with climate change? We know Russia is building up a military presence in the Arctic. Or do they want to pull more resources out of the Artic? Food for thought.

Of course, every nation will be impacted if there is a huge global warming crisis in the future. Let's keep our eyes and ears open and prepare for whatever happens. I will not be frightened into signing away my rights and freedoms because my government says they're all related to climate change.[157] So, there. I just want to end this section with an observation. Look at the title of this White House report I mentioned earlier. It starts with "Findings from Select Federal Reports." The operative word is *select*. Just saying. Are there perhaps other findings they left out because they didn't agree with the White House agenda? Thinking out loud.

(2) Climate change is a *threat multiplier*. I had to go look for a definition of threat multiplier.[158] A threat multiplier is anything that exacerbates what is going on. For the GND, it's climate change. Luckily, my source goes on to do some more explaining. I can now opine that poor old climate change is being used as a scapegoat for all our ills. No one in government has to take any responsibility for anything. It doesn't take a genius to spot the donkey in a flock of geese.

> Climate change is often viewed as a 'threat multiplier,' exacerbating threats caused by persistent poverty, weak institutions for resource management and conflict resolution, fault lines and history of mistrust between communities and nations, and adequate access to information or resources." (United Nations)[158]

That sounds more like political oration than a definition. Since the GND doesn't define *threat multiplier*, I'll have to go with my resources. It sounds like the GND is presenting the idea that whatever the planet does, climate change is at fault and making everything worse. The question is, do we really have this horrible climate change? Personally, I don't think we do. At least not yet. Using climate change as an excuse for all our ills excuses certain people and groups from any wrongdoing and blame.

It's climate change's fault the US has systemic injustices, economic injustices, and racism. I hope the rest of the world has the same because it wouldn't be right if climate change only affected the US this way. Severe climate change is upon us ... Look into my eyes ... Severe climate change is upon us. You must believe! Yeah, right! And I've got some horse tonic to sell you that will cure all *your* ailments. Besides, if we have threat multipliers, wouldn't it stand to reason we should have threat minimizers as well?[158] And indeed we do. According to a report by the UN General Assembly, there are 6 threat minimizers:[159]

Adaptation: Over this planet's existence, organisms and species have changed their structure and/or function to improve themselves to better survive in a changing planet. Humans have adapted since before they became bipedal. Humans have collected into communities to hunt and grow food to better survive. Then they began warring. We, as humans today, may need to adapt as well to a changing planet.[160] And we are still at war with one another.

Economic development: It seems obvious that more economic development would help mitigate any climate change. More lower income areas would have better opportunities to rise in the world, making sure that we can adapt to new environments. There have been many discussions on this topic. One idea states as the world economies grow, as well as population, so does the demand for resources, and that leads to more destruction of our environment.

I don't think that's what the GND has in mind. It probably espouses the idea that economic development in the world would raise the living standards of vulnerable communities, thus creating a population that can adapt to changing environmental conditions. The GND doesn't, however, mention the horrible losses to our environment caused by our expanding population and renewables.[161]

Governance: This is a short one. With the right governing power, policy makers can create a model of renewable resources for reaching net-zero. The right *governing power* means the far left-wing socialist politicians. They know what's best for the rest of us. Keep *us* in power while the rest of us pray for freedom. Phooey! Right now, it seems most likely their solutions will not work. They have created record high inflation. Not Putin. Gas prices are through the roof. Again, not Putin. So right now our *governance* is in the toilet.

Capacity building: We believe that the GNDers won't like this one even if it does come from the UN. Capacity building is about promoting *local* communities to develop their own solutions to problems. It helps communities shape their own destinies. It helps build local confidence and leadership, strengthens local individual skills, and promotes a shared vision of the future. It's being responsible for your future.[162] Within the GND, the central government controls everything.

Mitigation: The objective of mitigation is to have minimal human interference with the planet's climate systems but still slow climate change so ecosystems can adapt naturally to a changing environment. What it's not supposed to do is destroy the environment to mitigate climate change, which is what we are doing.

Ancient civilizations as well as the environment have adapted. Some have migrated, and some have fallen, partly due to environmental changes. We, as a planet, will have to learn to adapt and work on better defenses against the rise in temperature. If we can slow the rise in temperature without destroying our environment, we will have more time to adapt. So far, there is only a small change in the planet's surface temperature. One mitigation factor is to stop the deforestation around the world and replace those trees.[163] It's not the only thing, but we must do it, or we will eventually be a treeless planet. Like Easter Island, the people disappeared.

Conflict prevention: The world has been immersed in tons of conflicts since we walked on two feet. It's led to the displacement of communities caught up in their violent civil wars and global wars. Some say climate change is partly responsible for worldwide civil wars. Apparently, it's because climate change affects water resources, food availability, flooding, and desertification. Bull radishes! The civil wars are led by evil, selfish tribal leaders. They wage war on their own people, committing genocide to rid the areas of rival tribes and different ethnic groups. Why, when the US sends food and water to ease the suffering of the poor, do governments hold back this aid? It's not because of climate change. So, what do we do? We stick our heads in the sand and blame everything on climate change. So, what do we do? The proverbial solution is for all these countries of the world, with the help of the UN, to work together to stop and prevent these violent conflicts.[164, 159] They never do. Many of these countries have been fighting for centuries, and they won't stop just because we ask them to. Look at the US right now. Instead of uniting the country in a common cause, the Left is creating an atmosphere of hate, division, and racism.

You see what I mean! Let's just blame climate change for human greed and hate. That's easier than blaming criminal networks, tribal groups, and governments for grabbing more territory, exercising control over our religion, or governments and guerilla armies annihilating entire communities out of sheer hate. All for what? Money? Power? Thinking out loud again.

That's it for the six minimizing threats. The next section of the GND (1, 2, 3) are all social issues. Let's go to the House *resolved* to achieving the following. I'm only dealing with section (1) and (1A).

(1) "It is the duty of the Federal Government to create a Green New Deal—." That's a barrel of poppycock. It sure isn't their duty;

maybe it's the Left's great desire but not duty. Their duty is to the *people* and not what *they* think we need.

(A) "To achieve net-zero greenhouse gas emissions through fair and just transition for all communities and workers." This means everyone must make the transition to clean energy for their homes and workplace, such as every home going solar and being all electric. Communities must work to transition to clean energy for businesses, government offices, transportation, and so on.

The first thing that comes to mind is the abrupt halt to the XL Pipeline. Where's the "fair and just *transition*"? Between ten and eleven thousand people were suddenly out of a job, and more on the way ... right in the middle of a pandemic. Few to no jobs are to be had. And most of the renewables are made in other countries like China. So, where are all these unemployed pipeline workers going to transition to? In the renewable industry? What are we going to do? Where is the government to help in this transitioning?

Transitioning to our current choices of renewable energy solutions isn't going to happen anytime soon. Plus, as much as we may want to build these renewables, it will destroy the environment. Our socialist politicians tell the out-of-work pipeline workers to just go get a job in the renewable sector. Move your family, take a huge pay cut, be happy.

Sections (B) and (C) are social issues. Create new jobs and invest in infrastructure. Great. We have not created new jobs in the renewable sector, and we're six months into a new presidency. We are also haggling over what is infrastructure. Geez, it's like walking on a treadmill; you walk and walk but never go anywhere.

(D) To secure their list of things like clean water and healthy food is not an easy task. We have to stop companies from polluting our waterways, like the EPA in Flint, Michigan. We need to grow healthy organic food by not allowing deadly pesticides to be used, which the FDA again has allowed. We need to not import food that is contaminated with human waste and deadly pesticides. We never hear a word from the EPA or the FDA and their shenanigans when they should be protecting us.

Then there is the ambition to secure the climate and a sustainable environment. Secure the climate? Wow! How exactly? With renewables that help destroy the environment. Renewables that need coal and petroleum. How about a sustainable environment? The larger our planet's population gets, the less sustainable our environment is going to be. You can't sustain the environment. It's going to change over time with or without a rise in temperature. That's what the planet does. So much for that. I hope you all understand this.

Everything from (E) to (2) are social issues and meaningless drivel.

(2) Goals in A through E will require:
   (A) It mentions *community resiliency*. This we do have to deal with in terms of climate change. Resilient communities are able to handle disasters as *they* see fit. This helps to get back to normal faster and easier. I don't know what the authors of this resolution meant by community resilience, so again, I'll rely on my research.

Community resilience is a collaborative planning effort by the people of a community who put together planning teams to come up with solutions best for their community.[165] Orders should never be spit out by presidents, governors, or politicians. Each community knows what's best for it. Each community is different. So, must we do what the central government tells us, or do we follow this resolution and get together with our community leaders to determine what is best for the community? Just asking.

My last comments have to do with section (A). It will be as short as my hubby's underpants. Building resilience against climate-related disasters sounds great. But we don't know if certain weather events are even related to this climate change. How are we going to change hurricanes or tornadoes? We can't. It's part of our planet. The planet will always have category 5 hurricanes and huge tornadoes. Lowering CO_2 won't change that. It's like trying to change lead into gold. Alchemy, I say! Perhaps the GND is trying to say if we lower CO_2 emissions, we can affect the strength and number of storms and lower our temperatures or at least stop the rise. To achieve this, they say they need to "leverage funding." Leverage funding is a tricky way of saying higher taxes, folks! For everyone!

(B) Repairing and upgrading infrastructure in the US, including...
- (i) I love this. Remember when I said this would be an interesting section? The GND's plan is to eliminate pollution and GHG emissions as *technologically feasible*! Think about that for a minute. Float it around in your brain for a while. Come on. Down in the deep recesses. Done? It's pretty simple. Got it yet? The phrase *technologically feasible* gives the GNDers and politicians who promised to lower CO_2 emissions and eliminate pollution—wait for it—the out they need when their big, inflated plans fail. We're sorry. It wasn't technologically feasible.
- (ii) Guaranteeing universal access to clean water. They're going to do this by going to (iii). Why can't they clean up Flint, Michigan? Is that even on their radar still? Climate impact isn't going to clean up the area around Flint. They say that (iii) will clean everything up. Oh yeah, (iv) needs to be done as well. Everything will be OK if we also pass any infrastructure bill that they present. They can't even agree on what infrastructure is. Here we go.
- (iii) By reducing the risks posed by climate impacts, and ... How can you reduce the risk? What do they mean by climate impact? The effects of hurricanes and tornadoes? How can you lower the impact of a tornado? I guess the GNDers mean the need for lots of these renewables. Lower CO_2 emissions. Weather will become less of a

disaster. Even if we lower our emissions more and more, it takes hundreds or thousands of years for climate to react. And that is only if current climate is truly affected by our slight temperature rise. Is that supposed to guarantee clean water? Along with (iv) of course.

(iv) By ensuring the passage of any infrastructure bill to include climate change. So (iii) and (iv) are supposed to remedy our clean water issue. Really? Again, Congress is arguing over what infrastructure really is. And, believe me, it's not what we think infrastructure is. I can't believe the GNDers think this hogwash will give us clean water and that thinking Americans will swallow this bilge. States need to fine, big-time, companies that pollute our waterways. It needs to hurt. Clean up the EPA for what they did to Flint, Michigan! And (iii) and (iv) will never give us clean water. We're almost there, they say. We're working on it, they tell us.

(C) Just a few words. Meeting 100 percent of our energy needs "through clean, renewable and zero emission energy sources" *isn't going to happen anytime soon.* So, when exactly are we supposed to meet this 100 percent renewable mark? By 2030? Even government agencies don't believe we'll see this happening in the foreseeable future. We'll talk more about this in chapter 5. It's a pretty complicated section.

(i) Dramatically expand and upgrade our renewable energy sources. They're already trying to do that now. You'll read a lot on renewables in chapter 5, but you'll read even more in the final chapters. And …

(ii) Deploying new capacity. They really didn't explain this one at all. I hope they mean better capacity in renewables. Maybe they mean better efficiency because that they sorely need. Maybe they mean more renewables. Maybe, by capacity, they mean more higher-capacity wind turbines, like 3, 10, or 20 MW capacity turbines. But then what about solar and all the rest of the renewables? I suggest we ask the GNDers to define their terms, so we know what the heck they're talking about. It sounds great but says absolutely nothing. Maybe it's a well-guarded secret.

All the next sections (D, E, F, G, and H) are things the GND espouses as long as they are *technologically feasible*. There goes that term again. What do you suppose that really means? Does it mean we have to do this particular thing right now, but we don't have the technology to do it? Or maybe we'd like to do it, but we can't afford the technology. Or, worst-case scenario, lobbyists pay them not to do it, and this will be their way out. I'm not saying that's what's going on, but it's just something that we should think about. You know. Tit for tat.

(D) Build or upgrade to be energy-efficient and to ensure affordable access to electricity. I think we do need to be more energy efficient. So, why build renewables that have terrible efficiencies? I'm all for local communities being able to upgrade their older buildings. I hear that in California, there's a mandate that all new housing should be built all electric. Again, this is putting the cart before the horse. First, produce tons of green energy that can support that. Getting rid of fossil fuels to invigorate your environmental political base without having a backup is irresponsible and just plain nuts. Things like that will cause many brownouts. I might be wrong, but I heard our progressive governor of California imported oil, so we don't have the terrible brownouts we've had in the past. Solar and wind just didn't cut it.

(E) *Electrification*. This is strictly for informational purposes. What is electrification? Basically, it is making homes and businesses all electric. The main system for electrification is the heat pump. It moves air from one space to another using electricity. Sounds like the air-conditioners we have now. Not quite. First of all, they don't produce electricity. They use electricity to move both warm and cool air.[166] We only pay an electric bill. They say it's cheaper than fossil fuels. Sounds great. They say our bill will be around 50 percent lower than what we pay now.

The big secret is once we go *all* electric, there's no longer energy competition, and the huge monopolies can charge whatever they want. What are you going to do about it? Go back to natural gas? Sorry, it's gone. We'll start seeing our electric bills go higher and higher. Companies want their profits.

Let's talk about the raw materials for these heat pumps. They use iron castings (need coal), steel parts (need coal), aluminum tubing, and probably copper wiring. All heat pumps need fluid for energy transfer. They can use water, hydrocarbons, or ammonia. At least that's what my source says.[167] I think water sounds good. The other stuff is nasty. We could use water, but that would take a huge amount of water—water we may need to sustain the planet and a growing population.

Hydrocarbons are essentially oils (petroleum). I wouldn't say much about this except for two things. It's fossil fuel based and spills. Since the plan is to electrify all of the US and the world, we'll need massive amounts of hydrocarbons and/or ammonia. Hydrocarbons directly affect the environment, big-time, in the way of spills. After all, it comes from crude oil.[168]

Ammonia is slightly different. A tiny amount of ammonia occurs naturally on the planet, but in larger amounts, it can affect farming by producing lower yields. Ammonia vapors are toxic to livestock and other animals. Train derailments with ammonia on board cause a great deal of damage to the surrounding environment.[169] According to the GND, all this electrification will spur "massive growth" in clean manufacturing. Manufacturing will still emit GHGs. Better yet, it will spur a massive migration of businesses to other countries like China. And, here it comes again ... it will be done "as much as is technologically feasible." Did you see that coming?

(F) (G) The following sections are all about cleaning up pollution, lowering CO_2 emissions, access to healthy food (so let's not import unhealthy, contaminated foods; let's grow organic in the US), and transportation going electric, *as much as technologically feasible*

(i) Supporting family farms. Great idea. It just seems that big government leans toward the huge farm conglomerates, leaving family farms in the dust. My observation.

(ii) Invest in sustainable farming. What do they mean by sustainable? I discovered there is no good definition of sustainable farming. There's a lot of talk and yapping by media and the farming community itself as to what it means. The best thing I can come up with is we must set a goal to save our planet by providing the basic agricultural needs of its people. We must protect the land while providing food. At the same time, we need to balance our resources for future generations. How is that for a definition? At best, it's incomplete because it must adapt and change as our population increases. I think we began our sustainability discussions at the time of the Dust Bowl in the 1930s. You have to realize that sustainability programs need to change with changing demands. And I believe we are good at that.

(iii) Universal access to *healthy* food. If that's what you are calling for, then *don't import from countries that use unlimited pesticides and especially from China*.

And so on ... "*as much as is technologically feasible.*" (OK, I'm bonking my head on my computer in sheer insanity, laughing, cackling as if I was really crazy, but I'm not crazy.) Do you get the joke? Access to healthy foods, but not really because it's probably not feasible at this time. Our socialist politicians are crazy. The politicians who signed on and agree with the GND and this "access to healthy foods," go and import crap from other countries. Get your priorities straight, will you?

The rest of the GND is how government will do all this stuff for us but only if technologically feasible. Whatever they can't do becomes technologically unfeasible. But we're working on it. (H) and (I) aren't in the purview of this book. (H) talks about transforming the transportation system to a clean and affordable system. Go for it. Great. (I) talks about mitigating and managing the long-term effects of pollution and climate change. That's great too. Go for it. However, you know they're going to form committees to study the problem for years to come. Again, the GND and its authors don't know how they plan on doing any of this. Pass legislation for huge spending bills is their solution. I haven't seen any pollution cleanup by the government. You can't mitigate pollution. You have to stop the pollution and clean up the rest of it. They just pass legislation, and nothing gets done.

(J) This section looks at CO_2 from another angle. It talks about systems to pluck GHGs right out of the atmosphere. Pig pucky, you say! It's not as far-fetched an idea as you might think. We will need to remove billions of metric tons of CO_2e every year as well as reducing emissions. If this is going to work, we will need some new technologies. The question is, are they going to work? My research landed me right in the middle of this hot issue. I found articles on the ways to remove carbon from the atmosphere.[170] First, restore the forests ... Afforestation. This means globally. We should all know what photosynthesis is. Simply put, it removes CO_2 from the air and puts out oxygen. This doesn't mean we can just plant trees and it's all OK.

The hard part is making our forests healthy by good management practices so more CO_2 can be reclaimed. We are talking billions of trees worldwide. The problem with planting more trees is that it takes fifteen to twenty years or more to absorb the same amount of CO_2 for the trees you just cut down. Every acre of healthy forest can absorb three metric tons of CO_2 per year. An acre of baby trees can't even begin to equal that. Trees are a great idea because you will eventually get cleaner air and water. But planting trees can't be our only solution.

Let's move along to soil. Soil naturally stores carbon. The problem here is that governments would have to create incentives for farmers and landowners to capture more CO_2. Farm soil has a lot less carbon because soil is used and turned over so much for agriculture. It releases the stored carbon. Building soil carbon is not only good for the atmosphere but also for farmers and ranchers. Every small increase could really be helpful for a carbon capture program. Other programs like planting cover crops helps as well.

Direct carbon capture (DAC) is not a new idea, but it hasn't really caught on until recently. The problem is it needs to be powered "by low- or zero-carbon energy sources to result in net carbon removal."[170] Something that we would have difficulty producing right now. More tech development is needed in this field to really get it going. I've heard a lot more talk about carbon capture lately.

DAC chemically takes CO_2 directly out of the air, then stores it. However, it costs anywhere from $94 to $232 per metric ton. Another problem in this process is that it requires a lot of power. According to the World Resources Institute, "scrubbing one gigaton of carbon dioxide from the air would require about 7 percent of all projected U.S. energy production in 2050."[170]

Like the air capture system, there is seawater capture. It extracts CO_2 from seawater instead of the atmosphere. Seawater capture has several problems. It needs to work through its process often in bad ocean environments. There is also research going on by the US Navy, but again, it will be a useful tool in the future. Removing CO_2 from the oceans allows more CO_2 to be absorbed into the oceans from the atmosphere. Since seawater is heavier than air, this means that more energy is required for this process.

These last two sound kind of crazy. But they are being done. There are minerals that react with CO_2 that turn carbon from a gas to a solid form. It's called weathering.[170] This reaction happens in nature, but it takes place very slowly in geological time. Darn it! We know this happens naturally. All scientists have to do is figure out how to make it go a lot faster. It seems they are researching this very idea, and it can be done. It is being done. Very cool.

The last one I really don't like at all. It's using bioenergy with carbon capture and storage (BECCS) systems. We'll discuss bioenergy (biomass) for energy production in the next chapter on renewables. I dislike biomass energy. It burns trees, wood of any kind, bushes, and so on. It emits a lot of CO_2 and lots of other junk into the air. This idea is fraught with trouble. You'll understand why in the next chapter when we go into more detail on biomass.

Let's go back to weathering for a minute. I don't understand how it works, minerals that react with the gas form of CO_2 and turns the carbon into a solid. It sounds like a great idea. Researchers in Iceland succeeded in turning CO_2 into rock.[171] They have sped up a process that normally takes millions of years in nature. This technology is so new it may take a few years to get it usable on a large-scale basis.

Another process that researchers have discovered gobbles up carbon on the sea floor. It seems that benthic bacteria are absorbing carbon and using it in their own biomass. Scientists know very little about how it works, but it is being studied. Maybe it can be helpful in decarbonizing the oceans. It's worth investigating.[172]

Benthic microorganisms are part of microbial communities in lakes and oceans where sunlight reaches the sediment. The organisms use the energy from sunlight to eat up carbon and change it into matter through photosynthesis. All this trapping of carbon is wonderful. It's about as wonderful as serving hot chocolate in a thimble. We need to find ways to speed the processes up.

A new study by the University of Southern California, the Australian National University, and Lund University have looked at several areas of the ocean floor. It found an enormous source of carbon emissions that are not accounted for by our climate scientists. This carbon is not calculated into their climate models. This trapped carbon could be released when the ocean temperature reaches above a certain level. I don't know what that level is. This study even says that these gases could have gushed out thousands of years ago and possibly helped end the last ice age.[173] Something to ponder while sitting quietly by the fireplace on a cold winter's eve.

Going through sections (K), (L), (M), and (N), there's a long list of things we, as communities and countries, can work on. These are things our country can certainly work on, but the world has to do these as well. Not all listed here will do anything for temperature rise, but it's very good for the health of the ecosystems.

- Restore and protect threatened ecosystems.
- Clean existing hazardous waste sites.
- Identify and create solutions for other emissions and pollution.
- Promote international exchange.
- Develop the GND.

This section is a list of great things to do (except the GND) by local, state, and federal agencies. Some of it is being done by independent groups like the Sierra Club, 4Ocean, and many other groups. Each country should follow this list to make this planet a better place. Many countries will do nothing. Remember, I am only saying the global temperature has risen a minimal amount. The US should clean up its mess but not at the expense of losing thousands of jobs. It's going to be an extremely slow process because of the political games going on. There's a lot of gum flapping with few results and that includes you Republicans.

This is where I'm going to have to talk slowly, think quickly, and ride a fast horse. *What about China?* I'm going to stick to my personal opinion here, then ride out really fast. Is China going to want to work on temperature rise? Or climate change? They tell everyone they want to (My cat just hacked up a hairball). My observation: China wants our jobs, our technology, our resources, our money, and our country. I read that China recently built three times the number of coal plants than the rest of the world combined. Boy, they're working on climate change, aren't they. My horse (and cat) and I are now riding away really, really fast.

(3) The GND must be developed. No, it doesn't.
(4) To achieve the GND will require the following:
   (A) Sounds like the government has to *leverage* (force or trick) the people into accepting the GND and then tax us from here to hell. The rest of Resolution 109 is about mobilizing the GND, making laws (restricting the people), helping vulnerable communities, diversifying industry (send it to China and other countries), and a whole lot of social issues that take us to the end of this resolution.

By now, your head must be turning to mush. Have no fear, truth is here. I find it amazing that if we fix climate change, everything in the world will be right. But let's not spend any more time on the fantasies of the Left progressives and get back to what we are here for. In the next chapter, we will discuss the drive for renewable energy.

All this offends my sense of Peace.
—Hopalong Cassidy

5

Renewables, Anyone?

We will devote ourselves to discussing renewable energy sources as prescribed by the GND and show how these are ineffective, inefficient, and destructive to the environment.

Let's get down to the nitty-gritty of renewables. Remember, the GND demands that we get to 100 percent of our US energy needs through "clean, renewable, and zero-emission energy sources." Is that even achievable? In this chapter, we'll run through the different kinds of renewables. You'll find out how efficient our renewables really are. We're going to do some calculations to show that the GND renewables are very inefficient and ineffective for the world's needs and would require massive land use, which would be environmentally destructive. You're going to see how much mining, processing, steel, copper, and so on is required to build our renewables, as well as the environmental devastation they cause.

Wind Turbines

There are oodles of components to wind turbines. There are many kinds of turbines. Energy production by wind is measured in kilowatts (kw), megawatts (MW), or kilowatt hours (kwh) or (Mwh). Turbines are rated by how many megawatts per year they are *supposed* to produce.

A 1 MW capacity means it is supposed to produce 1 MW of energy per year or 1,000kw. There are many others rated at 1.5, 2.0, 3.0 MW, and even gigawatts (GW). There are plans to build one as tall as the Eiffel Tower. Eee-gads! Can you imagine that thing self-destructing? Just be aware, the taller the tower, the longer the blades, the more land is required. The larger the component parts, the more steel, copper, concrete, fiberglass, and more are needed. Thus, more mining.

Starting this section was about as hard as catching a piglet in a foot of slimy mud. I've spent so much energy getting this wind power under control, I about gave up this whole book idea. But let's press on. We need to get this section as complete as possible so the following sections can be better understood. We'll be doing lots of calculations to show how much electrical output turbines actually produce. OK, wind power. Here we come.

1. Electrical Output

When discussing the energy production of wind turbines, it will be in terms of the energy produced per year. For that, we need to know how wind turbines work. In a nutshell, the wind comes along and turns the large blades (and I mean colossal). As the blades turn, so turns a rotor inside the housing (nacelle), and it turns a shaft. That shaft powers an electrical generator. Simple concept. Here's the catch. Just because a turbine is rated at 1 MW/year or 3 MW/year capacity, it doesn't mean that's what it actually produces. Got it so far? Because of physical stresses like friction, downtime, and wind variability, *turbines are not 100 percent efficient*, so now they are given an *efficiency factor* (EF) or power output. It's usually somewhere around 30–50 percent, give or take. Other experts have determined the efficiency factor to be 15–30 percent as more typical. So, science does what it normally does: it gives us averages. Let's see how we can calculate the efficiency factor (power).

Power = air density (kg per meters cubed) x swept area of blades (meters squared) x wind speed cubed (meters per second) ÷ 2 [58]

This gives the efficiency factor for each individual turbine. The calculation has already been done for us, so we don't have to do that one. Thank heavens! The turbine 200 feet away may have a slightly different EF (efficiency factor). If we *assume* a turbine runs 365 days a year and twenty-four hours a day, with no friction (which is impossible), we can calculate how many MWh/year that one turbine *should* produce. That's great information for the wind farm.

Many wind scientists, when performing their calculations, like to assume that turbines run 365 days per year and twenty-four hours per day and use these numbers in all their calculations. They are giving misinformation. Since each turbine is slightly different, we need to determine the averages and work with that. It's something scientists do all the time.

We know turbines *don't* run 365 days per year (days/year) or twenty-four hours per day (hours/day). Haven't you seen wind turbines stopped or turning very slowly as you've driven around? And that's what you'll see translated in the following calculation section.

What is the efficiency factor based on? Efficiency is the amount of kinetic energy in the wind that is changed to mechanical energy then electrical energy. What science does is change the kinetic energy to mechanical energy to electrical via mathematics. Since many of you may not want to wade through calculations, they will be in appendix I—Wind Turbines. The results will be posted here.

There are several steps we need to climb to reach our final answer of how much energy a wind turbine actually produces per year. We need to know the following:

1. Capacity rating of the turbine
2. The efficiency factor
3. The *average* number of days a turbine runs per year
4. The *average* number of hours a turbine runs per day
5. How to convert MWhrs/year to MW/year

Formula 1: Calculate the megawatt hours (MWh) per year a turbine actually produces:

The average number of days/year a turbine is running and producing power (times) the average number of hours/day it runs (times) the capacity of the turbine (times) the efficiency factor equals the MWh/year it actually produces. We will be using a 2.5 MW–rated capacity turbine[59] because as of 2019, the US average capacity for a turbine was 2.43 MW.[60] Remember, every turbine is slightly different in their capacity, blade size, and height. I can't measure every single turbine in the country, so I'll go with the *averages* I find in the data.

Formula 1:
days/year X # hours/day X Capacity X Efficiency Factor = MWh/year

Appendix I—Wind Turbines: This shows the numbers so you can see how the energy output goes down as the efficiency factor goes down. The average efficiency factor is 35 percent.[59] And that is being generous.

See appendix II—Calculations for Wind Turbines.

I hope we can keep all these numbers straight. We don't want to confuse ourselves, but we can see that the MWh only takes a nosedive because no turbine runs 365 days per year or twenty-four hours per day. Turbines do not run at optimal speed all the time (winds around 27–56 mph). At slower speeds, the production falls off. No wind means no production. When the winds are over 88 mph, the turbines must shut down or risk potential damage. If they don't shut down, they might even explode or fall apart. Turbines are also shut down for maintenance.

The last thing the industry and politicians would like us to know is that they may shut down turbines because the grid has too much power and less demand. Really! I mean really! When the demand for energy goes down, the renewables get shut down first. That's not exactly the whole story. The grid gets its power from many sources—coal, nuclear, hydro, solar, wind, and so on. But when demand goes down, then one of these energy sources needs to be slowed down or temporarily shut down. It is way too difficult to slow or stop coal or nuclear plants, but renewables like wind and solar are much easier to stop and restart.

Let's see what happens to the MWh when we change the number of days in our formula. I feel generous to the plight of the wind turbine at the moment.

Formula 2: Change the number of days a turbine runs per year (in the formula).
See appendix I—Wind Turbines.

Using 245 days, which *is* the average number of days per year a turbine runs, a 2.5 MW capacity turbine produces per year = 5,145 MWh/year.

We also need the average number of hours a turbine runs per day. That's coming up. It turns out it's 16.4 hours per day.

You'll see how production of energy goes way down when a turbine doesn't run or runs less of the time. Next, I'm going to convert these MWh back to MW.[61] Why? Because this number has to be the same units as the rating of the tower, 2.5 MW. We can compare MW to MW.

Formula 3: Convert MWh to MW.
MW = MWh/year divided by number of hours run/year
See appendix III.

Formula 4: Hours run per year and per day.
Average number of hours run per year = 6,000 hours/year
Average number of hours run per day = 16.4 hours/day[62]
See appendix II.

The calculated MWh for a 2.5 MW turbine = 3,515 MWh/year
See appendix III for calculations.

Formula 5: We can now convert MWh/year to MW.
A 2.5 MW capacity turbine produces around 0.60 MW of electricity a year, *not* 2.5 MW.
Here is one comparison:
One piece of scientific data I found said a 3 MW tower can produce

7,884 MWh/year. So, at 30 percent efficiency, if we take our formula from earlier:

365d/year X 24hrs/day X 3 X 0.3 = 7,884 MWh/year [63]

Again, that's assuming it runs 365 days a year and twenty-four hours a day ... which they don't. Another answer I found said a 2.5-3 MW tower produces about 6,000 MWh/year.[223] I used these to show how officials can manipulate numbers. Each of these calculations uses 365 days a year and twenty-four hours/day. My own research found that the *average* number of days and hours any given turbine runs is 245 days/year and 16.4 hours/day.

It seems that the pro-turbine officials have calculated higher numbers than we have because they are using 365 and twenty-four, which turbines *don't* run because they are not 100 percent efficient. I hope you all have been able to follow this because we now have to go back to our 2.5 MW turbine.

Hope we've kept our math straight. In short, a 2.5 MW turbine might only produce (on average) 0.6 MW/year.[63] Over, let's say, a twenty-year lifetime for a wind turbine, 20 years X 0.6 MW/year = 12 MW = 12,000 kw over twenty years. That's pretty dismal.

2. Construction and CO^2e Emissions

This section has a lot of parts to it. First, dimensions of a turbine. We are using the GE (General Electric) 2.5 MW turbine:

144 feet blade length (one blade)
297 feet hub height
423 feet total height, including blade height
1.5 acres rpm range [64]

The major elements for a 2.5 MW wind turbine are (best estimates):

- 309+ tons Steel (a 2 MW tower averages 309 tons steel)[65]
- 45+ tons steel rebar[66]

- 120+ tons cement to make the concrete
- 1,200 tons concrete[66]
- plus 20,000 tons concrete for miles of new roads[67]
- 8+ tons of copper[68, 69]
- 2,000-pound (1 ton) generator
- 6 tons wiring, tubing, cables
- 2,000-pound (1 ton)+ step-up transformer
- 36 tons+ fiberglass with resin blades[70]
- hub and nacelle; 14 tons plastics, 800 pounds neodymium, and 130 pounds dysprosium large magnet (MIT study shows 2 MW turbine has 752 pounds rare earth minerals)[71]

And that's for a small turbine. A larger turbine will use much more. I'm going to jump right in the stew pot and go through how companies make this stuff, so you see the CO_2e emissions and dirty by-products that go into the manufacturing.

1. Steel

Before making the steel, you need to make pig iron. Pig iron can't be mined. It has to be made from iron ore, coke, and limestone. Coke, as well, can't be mined but has to be made using coal.

The iron ore and other materials go into a blast furnace called a stack. The top part of the furnace produces gas from the burning coke, which removes a lot of oxygen from the iron ore. About halfway down, limestone starts to react with impurities, which forms a slag that floats on the surface of the molten pig iron. The slag is drawn off, and the molten pig iron is tapped off. The process begins again.[72]

The molten pig iron goes through a final process in making steel, and that's blowing oxygen at supersonic speeds through the molten pig iron. This lowers the carbon content of the pig iron and changes it into low-carbon steel. Limestone gets added to help remove the impurities. Lastly, they add stuff to make it stronger, flexible, and so on.

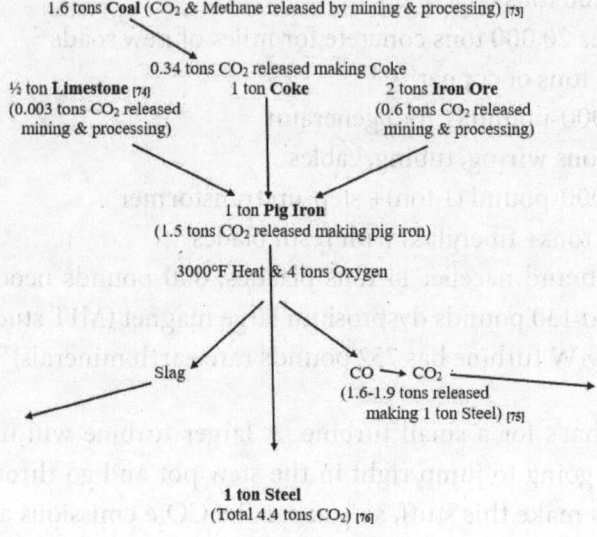

Fig 9 Steel Production

Steel production

To make steel, you need to start with coal. Coal mining, in itself, produces little CO_2. Its major GHG is methane, and that they are trying to capture while mining. Most mines do a so-so job. So, why the big fuss over coal? It's the burning of coal that's the problem. Combustion produces tons of CO_2. We're not talking about combustion of coal when dealing with the steel-making process. At the moment, we are only talking about coal mining. The good thing about methane is most of it, in the US and several other countries, gets recaptured to use somewhere else. There's a big push now to do something about methane.

It's difficult to determine just how much methane is released because there are so many factors involved. It depends on whether it's underground mining or surface mining, whether methane is recaptured or not, the ranking of the coal, and the depth of the coal. You got the idea? I tried to figure things out with the data I could find. It's very confusing. Even when methane is recaptured, some of it still gets out. How much? Who knows? There are a bunch of calculations for figuring out methane (CH_4) emissions, but I'll try to simplify it.

There are tons of papers out there about methane and coal mining. I mostly had to convert units like cubic meters (m^3) to tons. According to the EPA, the 2020 estimates for *global* methane emissions from coal mining reached 800 million metric tons of CO_2e (MMT CO_2e) or 881 million tons. CO_2e in this case is mostly methane. The worst offender is China, with 420 MMT CO_2e for 2020. That translates to 462 million tons. The US produces around 80 MMT CO_2e or 88 million tons of non-CO_2 emissions. These were predictions for 2020 on past and current data.[77]

These figures are for working coal mines. There is also methane release from abandoned mines, post-mining work, and surface mines.

I don't believe they account for the methane that is recaptured. In the US, most of the methane is recaptured as far as I can tell. Here are the CO_2 emissions for the turbine and they are minimum numbers.

4.4 tons of CO^2 per ton of steel
They need 309+ tons steel for a 2.5 MW tower = 1,360+ tons CO^2
45 tons steel rebar X 4.4 tons CO^2/ton of steel =198+ tons CO^2

2. Cement and Concrete

Let's take a look and see how big a wind turbine is.

We are using the GE's 2.5 MW wind turbine.

First, we're going to take a look at cement. Cement is the base for making concrete. Cement is made through a chemical reaction of calcium carbonate (limestone), silicon, aluminum, iron, and smaller amounts of other stuff.[78] I'm not going to go through the entire process like I did with steel. I'll go straight for the numbers my research found.

0.9 pounds of CO_2 is produced per one pound of cement.[79] It's almost a 1:1 ratio, so let's just call it one pound CO_2 for one pound cement. One ton of cement produces one ton CO_2. (For reference, two thousand pounds equals one ton.)

Don't forget that about 20,000 ± tons of cement is needed for the roads that have to be laid down for all those huge trucks and cranes. The bigger the wind farm, the more cement is needed.

The CO_2 is produced in two different ways. It uses fossil fuels in the burning process and calcination, which is when calcium carbonate is heated and breaks down to calcium oxide and CO_2. Concrete is cement with some kind of aggregate in it—usually small rocks, but it can be almost anything. One cubic yard (4,060 pounds or 2.03 tons) of concrete emits almost 400 pounds (0.2 tons) CO_2. So 2.03 tons (t) concrete emits 0.2 tons (t) CO_2. The good news is the CO_2 produced in manufacturing cement is partly reabsorbed back into the concrete through carbonation during its life cycle.

1,200 tons concrete needed: $0.2t\ CO_2/2.03t_{concrete} = x\ tons\ CO_2/1200t_{concrete}$
2.03 tons $\mathbf{X}$ x tons CO_2 = 1,200 tons $\mathbf{X}$ 0.2 tons CO_2
$x = \underline{1,200 \times 0.2} = 118+$ tons CO_2
      2.03

We can't forget the emissions to make cement ...

One ton of cement produced one ton CO_2. The cement needed is 1,200 tons minus the aggregate. I don't know what that is. So, a little funny math, and I figure half of the 1,200 tons concrete we can attribute to cement. It might be different for different projects.

600 tons cement = 600 tons CO_2

So, for the cement and concrete = 718 tons CO_2 emissions for a 2.5 MW tower.

What is the life cycle of concrete? According to ResearchGate and scientists at Ghent University, the life cycle for concrete is about one hundred years, and this is the minimum life span.[80] After its use is over, concrete can be recycled. But then that takes CO_2 emissions to recycle and use it for something else.

3. Copper

Mining copper has a lot of variables. Determining the CO_2e depends on a lot of things: the grade of the ore, how deep it is, mineral properties, and where it is. Copper is found in the earth's crust and the oceans. It comes in the form of an ore, so the copper has to be extracted from the ore.[81] I won't go into any great detail. There are two processes to purify the ore. They are electrolyte refining and leaching.[81]

How do we figure out how much CO_2e is produced? Again, we have to take the averages. After all is said and done, taking everything into account (not the mining equipment), the average for copper is 5.5–9.5 tons CO_2e per ton of copper.[81] So, I'll take the average of these numbers, which turns out to be 7.5 tons CO_2e per ton of copper.[81] *It does not include all of the mining equipment.*

Copper is very important to human living, but it doesn't come without an environmental price tag. There is significant and sometimes permanent ecological damage. If not controlled, the damage to the groundwater, ecosystems, and surface water can be catastrophic. One US mine produced 110 million tons of tailings rich in sulfuric acid. They also contain arsenic and other heavy metals, and as the tailings erode, they leach into the groundwater, lakes, and rivers. Unless hundreds of millions of dollars are put in for remediation, mines should be liable for damage beyond their borders.[82]

There's more bad news. Certain minerals called ore bodies are found right along with the copper, and they contain radioactive materials. The two that strike me as important are uranium and radium. Copper mining waste in the US makes up the largest percentage of waste of any metal-mining process. There is a huge amount of waste in copper mining, including the acids, uranium, and radium, all of which have the potential to leach into our environment. The amount of copper extracted is small compared to the total amount of rock mined. The waste piles could exceed one thousand acres or more. Several hundred tons of ore must be processed for every ton of copper produced.[83]

It's no wonder environmentalists want to shut down copper mines.

They wreak havoc on the environment. The bottom line is, without copper, there is no electricity. Unless we want to use gold. However, there are many other uses for copper besides copper wiring. What else do we use copper for? Pipes, computer heat sinks, microwaves, ovens, vacuum tubes, ship building and more.

4. Nacelle

The nacelle is the housing that holds a massive generator and all the moving parts that turn the mechanical energy of the blades into electrical energy. As the blades turn, so turns a rotor to a faster speed, which, in turn magically becomes electricity in the generator. The nacelle alone can weigh fifty-six tons for just a 1.5 MW tower. A larger turbine means a bigger and heavier nacelle.[84]

The generator needs strong magnets to create a magnetic field in order to generate electricity. There are two ways to accomplish this. The first is to use permanent magnets from the mining of rare earth metals. Neodymium, praseodymium, and dysprosium are the elements used. Nasty mining no matter how you look at it. Plus, China holds between 80 and 95 percent of the rare earth elements (REE) and does the processing. So, the world does not want to piss off China. Right? The US was looking at mining our own rare earth elements. Oh, but I'm mistaken. Not anymore. Before our 2020 election, the democratic presidential candidate vowed to produce it here in the US. But after the election, the new democratic president announced we would import these rare earth elements. I guess from China.

Going on. The second method is using conventional magnets with copper windings as a collector. Again, nasty mining that copper is. The REE used right now are around 441 pounds/MW. Wind turbines use both the gearbox and direct drive. Either way, as turbines get larger and larger, they are going to need larger rare earth magnets or lots more copper. That means more mining.[85]

5. Fiberglass and Resin

Wind turbines use fiberglass in the blades and for the nacelle cover. A single 144-foot blade can weigh six tons+.[86] So three blades weigh around eighteen tons. With the hub, the whole mechanism can weigh more than seventy-four tons. I'm going to try to give you the short version of how blades are made.

Some newer blades are *composites*. These are materials that have different chemical and physical properties. When you put them together, they don't completely mix, but they are stronger. For turbines, they use "resins of glass fiber reinforced polyester, glass fiber reinforced epoxy, and carbon fiber reinforced epoxy."[87]

The manufacturing process involves both high-tech digital operations and manual labor. I think it's important to understand what's in some of our blades. You've got the polyester or epoxy. You form the blades. Then cover the blades with gallons of gooey, dark resin. Once the resin has cured, we're off to the next step.

The blades are in two halves at this point. The workers glue another piece in the center for structural integrity. They slowly close the two halves together like a sandwich. The last step is adding a special paint that helps protect the blades from dust, water, and wind. Then they add the flange and all the nuts and bolts to the end of the blade so it can be attached to the rotor at the proper time.[88] That's about as short as it gets.

How to determine the carbon footprint? There are so many steps and products that go into making just the blades and other pieces, it's very hard to get anywhere near a factual number. Every factory that makes the glass fibers, the resins, polymers, balsa, and so on has its own carbon emissions data. But I did find a chart of emission factors for different products, which will help figure out the emissions. For fiberglass, it's 2.6 kg CO_2e per kg of fiberglass. Here we go with calculations. Approximately eighteen tons of blade: if 1 kg of fiberglass = 2.6 kg CO_2e then 907 kg = 1 ton CO_2e, then 907 kg x 2.6 kg CO_2e = 2,358 kg CO_2e = 2.6 tons CO_2e for one ton of fiberglass. But there are approximately forty-five tons of fiberglass in a wind turbine. And 2.6 tons CO_2e/ton fiberglass x 45 tons = 131 tons CO_2e for fiberglass approximately. This includes the processing of the materials, not the transportation, assembly, machinery, or the specialized factories that have to be built.[89]

The important part of the composite processing is the terrible impact it has on the environment.[90] Besides the processing of the materials, the final product like the blades, hubs, and nacelles are not fully recyclable at this point. They are working on it. You also need to consider the huge factories that are needed to make the things just for wind turbines. They all have their own carbon footprint. There are generally two options to get rid of these products: burning (oh, yuck; what does that do to the environment?) and sending them to landfills. Neither of these are very good options. This doesn't look so good for the wind turbine industry thus far. The blades are huge, really huge and extremely heavy. It takes specially built truck trailers, cranes, and rail cars to move them.

Several companies are working on trying to recycle the fiberglass. The problem starts right at the collection point. Companies come and cut up the blades on-site using regular trucks. Ah, more fossil fuels for the trucks and chain saws.

These composites are *permanently* cured and can't be remelted. They need to go through chemical and thermal processes. Mix this around with some other stuff to make a kind of thermoplastic pellet, which could be used for 3-D printing and other products. However, making thermoplastic pellets uses even more power and emits more CO_2e emissions. There is a lot going into landfills. This technology is still new, and there is a lot of waste. Turbine farm owners want the old blades to be removed right away. Sometimes it doesn't happen right away, which makes owners none too happy.[91] Sad to say, a lot of the fiberglass remains piled up in trash heaps. But they are trying.

6. Neodymium Magnets

We've gone over a lot of needs of a wind turbine, but there are a few more very nasty ones. Get your radioactive gear on because this one is a doozy. Maybe the politicians don't want to know and don't want us to know about mining rare earth elements that we need for the magnets in wind turbines. The term *clean energy* is going to sound like complete tomfoolery after we get through with this section.

Neodymium is part of a complex system of seventeen rare earth elements that are all found together in the earth. You can't mine them like gold or silver. You must use different processes to separate them from each other. Neodymium is the most used of the rare earth metals.

Mining and processing these metals generates toxic and radioactive waste of gargantuan proportions. That's why China is about the only country who does this on a large scale. First, you have to remove the earth and crush it. Then it is trucked to a plant that processes it into a powder and leaches it with toxic acids to remove impurities.[92]

The last part of the process involves separating each of the seventeen elements. China buys the world's powder and processes it into its individual elements. I won't go through all seventeen, but I will discuss neodymium. This final stage involves many steps, including what's called floatation, leaching with hydrochloric acid (HCl), solvent extraction, sodium hydroxide and heat, and a few more.[93] In all these processes, there is a huge GHG potential and great environmental damage.

We are going to start with the neodymium oxide. The first step releases carbon monoxide (CO) and CO_2. As the process continues, it forms PFCs like CF_4 and C_2F_6. There can be filters for all this, but PFCs can't be filtered. However, they are working on processes to lower these PFC emissions. We'll see if China will lower the PFCs. For now, estimations of these emissions are 7 percent CF_4 and 0.7 percent C_2F_4 with about twenty million tons CO_2e per year for thirty thousand tons neodymium per year. That produces 1,400,000 tons CF_4, 140,000 tons C_2F_4. The rest is CO and CO_2. That's a lot of GHG. The more we produce, the more GHG is emitted. The last thing (where have I heard that before?) is please remember these are estimations. This technology is being studied as I write.[94]

7. Lubrication

Drilling for Oil

Like any piece of sophisticated and expensive equipment, wind turbines need to be lubricated. There are many, many moving parts that need to be lubricated with different kinds of greases, fluids, and hydraulic oils. One item is the gearbox. On the larger turbines, they use around sixty gallons. Now that there are even larger turbines being built, even more oil will be needed.[95]

The gearbox is only one of the things that needs tender, loving care. Here are just a few things that need work: the generator, several points on the blades like bearings, the main shaft bearing, and the drivetrain. Turbines also use hydraulic oils in the braking system. With all these complicated systems, there are several types of oils, grease, and hydraulic oils required.[96]

What about all this oil and grease? Where do you think it comes from? Hold onto your pink shorts. All the oil and grease come from oil drilling. Another fossil fuel requirement for wind turbines. I'll go through some of the numbers for you. Let's start with a barrel of oil. Different oils weigh different amounts, so the barrels will weigh different amounts. I'm going to use crude oil.

A standard-size barrel holds forty-two gallons.

Crude oil weighs about 275 pounds a barrel.

To get one ton of oil, you need seven barrels of oil.[97]

These are the basic facts when it comes to measuring oil. It's not easy to figure out exactly what the CO_2e emissions are for oil production. First, there are different qualities of oil; different regions emit different amounts of GHG.

The Carnegie Endowment for International Peace put out an analysis called *Know Your Oil*. It analyzes a lot of things and has amazing data. The first thing it states is that CO_2e emissions are very difficult to pinpoint because all the different oil fields have different emissions. They, however, like a rooster counting his hens, did come up with some numbers. They tested thirty oils, and the GHG emissions ranged from 450 to 820 kilograms CO_2e per barrel of oil. That's a big difference. Again, with the averages.[98]

The average being 570 kilograms CO_2e/barrel oil = 0.6 tons CO_2e/barrel oil.[98]

That doesn't look so bad. Well, that just dills my pickle! Oh wait, that's just for one barrel. In 2018, US production of oil was 11 million barrels per day. In 2019, the US produced 12.2 million barrels (bbl) per day. Since I have more data for 2018, I'll use that for my calculations. Just a little set of easy calculations—and voila!

11 mil bbl/day **X** 0.6t CO2e/bbl = 6.6 mil tons CO_2e/day
6.6 mil tons/day **X** 365 d/year = 2.4 billion tons CO_2 per year

Yikes! That sure gives it a whole new perspective on oil. Mind you, this isn't all for wind turbines.

Refining Oil

That was depressing. Now that we have our barrels of oil, they need refining into the different applications. I'm going to just quote this article from the EPA on what refineries do. They produce so many things, including CO_2e.

> Produce, gasoline, gasoline blending stocks, naphtha, kerosene, distillate fuel oils, residual fuel oils, *lubricants*, or asphalt ... by the distillation of petroleum or the re-distillation, cracking, or reforming of unfinished petroleum derivatives ...

It goes on to discuss the GHG emissions as well:

> Emissions from venting, flares, and fugitive leaks from equipment (e.g., valves, flanges, pumps). In addition to emissions from petroleum refining processes, the

sector includes combustion emissions from stationary combustion units located at these facilities ...[99]

There are also emissions from waste landfills and wastewater treatment, petrochemical manufacturing, and so on. For 2018 (for 140 facilities reporting), emissions were 180.6 million metric tons of CO_2e or about 200 million tons CO_2e.[100]

I'll break this down to separate gases:

200 million tons CO_2e:
   197 million tons CO_2
   992,080 tons methane
   551,156 tons N_2O (nitrous oxide)[99]

I hope you're not as confused as a fart in a fan factory. All these numbers and calculations confuse me sometimes. I know these are huge emissions when we only need a tiny fraction for the oils and grease to maintain a wind turbine (but we need them), and there are so many more things we use petroleum for. I was going to wait until later to get to them, but now is as good time as any to bring them up.

Here we are in the wonderful world of *plastic*. The world went from zero in 1950 to 322 million tons of plastic in 2015.[101] Plastics don't just come in one size fits all. There is polyethylene and polypropylene, polyurethane and Teflon. There are reasons we went to these plastics. They can be produced with cheap natural gas. They are the lightest items produced; they float. Lastly, they resist water, seawater, solvents, heat, and sun.

These things are so great, but with all good things comes the bad. They won't degrade for centuries. I didn't realize this one, but as these plastics bounce around the oceans, scraping along the rocks and coral, they create microparticles that are then eaten by fish, then animals and humans. We are all eating microplastic. Once it gets into the food chain, it's everywhere. Let's put this in perspective. All the people who think it's fine to dump their trash on the beach and into the ocean will eventually end up eating the microparticles. You can't escape it.

There are more than six thousand items that use some form of petroleum, and unless we are willing to give these up, petroleum is here to stay. I'm not going to list all of them; however, I will list the ones I think our young millennials and generation X can't live without. One barrel of oil, forty-two gallons, makes 19.4 gallons of gasoline. That's a base to start with. Petroleum is used mostly for fuel—that is, gasoline, jet fuel, heating, and electricity. The rest is used for our everyday products.[102] Plastic is made from petroleum and is used everywhere: cars, toys, computers, cell phones, in our houses, and even clothing. Asphalt is a petroleum product used on our roads, and it's in our tires. Paraffin wax, for those who like to make candles.

Packing material	Surfboards	Paints
Ballpoint pens	Bicycle tires	Nail polish
Perfumes	Dishwashers	Helmets
Caulking	Transparent tape	CD players
Washers	Refrigerators	Umbrellas
Ink	Shoe polish	Soaps
Toilet seats	Deodorant	Toothpaste
Nylon rope	Hand lotion	Detergents
Telephones	Movie film	Cameras
Vaporizers	Dishes	Fan belts
Floor wax	Shampoo	Drinking cups
Cell phones	Car parts	Bicycles

And so on ...[103]

8. SF_6—Sulfur Hexafluoride (Dirty Little Secret)

Last, but not least, is SF_6. Sounds like a spy friend of 007. I don't know how much of this is still being used, but it appears worldwide levels are rising. This is used in the electrical sector of turbines to prevent short circuits, which can cause fires. SF_6 has the highest global warming potential (GWP) of all GHG. It is 23,500 times worse than CO_2. It does tend to leak, and that's how most of it gets into the atmosphere. We can't detect it because it is colorless and odorless. Special equipment is needed.

A study done in the UK found SF_6 was increasing by 30–40 tons a year. Remember, it is 23,500 times worse than CO_2. SF_6 carries quite a wallop. Another downside is that it can't be absorbed or destroyed *naturally*. We will have to find a way to replace it (which I believe they are working on it) and then destroy it.[104]

Each turbine normally uses about five kilograms of SF_6 per maintanence. That's not so much, is it? However, that is equivalent to 129 tons CO_2. Engineers have come up with a vacuum technology that helps, but we still need alternatives to SF_6 that are economical and easy to transition to and that work.

9. Building a 2.5 MW Turbine

My approximate numbers for building one 2.5 MW turbine:

> 309+ tons steel = 1,360 tons CO_2 [76]
> 45+ tons rebar = 198 tons CO_2 [76]
> cement (10% of concrete) = 120 tons = 120 tons CO_2
> 1,200 tons concrete = 123 tons CO_2 (for the tower base)[76]
> 8 tons copper = 7.5 tons CO_2e/t Cu = 60+ tons CO_2e
> 45+ tons fiberglass and resin (blades, hub) = 131 tons CO_2e[76]
> 56+ tons nacelle (fiberglass exterior and interior components)
> > 25 tons (estimate) fiberglass = 72 tons CO_2
> > 36 tons steel (estimate) = 158 tons CO_2
>
> 2.7 tons neodymium magnet [76] = 1,800 tons CO_2e (126 tons CF_4 and 12.6 tons C_2F_4)
> 1 ton oil for maintenance
> 5 kg SF_6 = 129 tons CO_2
>
> Total tons = 4,152+ tons CO_2e.

And that's on the very low side.

Average land requirement: fifty acres per MW of capacity. Straggled along a mountain ridge, turbines are around 10 MW per mile. [84]

A 2.5 MW tower needs 0.25 mi^2. They need the distance to maximize the wind power. They also need to cut down large swaths of forests to maximize wind flow. What do you think about that environmental damage? Some experts say that the land under the towers can be used for cattle grazing or farming. Then again, others say you can use it if you want to risk your life. Most wind farms have restricted access even to the landowner. Just as a sidenote, when the weather gets cold and ice builds up on the blades, the blades could very well break. Just saying.

We are not quite done yet. These figures only include the CO_2 or CO_2e from a mine or actual tons for pieces manufactured for a turbine. Not included is the building of the factories, processing plants, use of mining equipment, and fuel. I'll see what I can do to highlight the important factors.

I'll start with mining. Surface mining uses a lot of energy as well as underground mining. That goes for pretty much all mining (coal, metal, and nonmetal). For the wind turbine, this includes mining for iron ore, limestone, silica (silicon), bauxite (aluminum-based ore), rare earth metals, coal, copper, and so on. This sector consumes about *365 billion kWh/year = 42 million MW/year.* This is strictly for the operations part.

Just picture explosives blowing huge holes in the earth to free the ore. Then, like wild dogs tearing apart their prey, the huge bulldozers, cranes, and trucks come in to grab the ore. That means diesel fuel, "electricity, natural gas, coal and gasoline."[105] Mining equipment like the bulldozers, hauling trucks, shovels, wheel loaders, conveyer belts, crushers, and lights consume energy. Lights for night work, lights and fans for the interior of mines, and there is electricity consumed by electric shovels and more and more and more. All this adds to the carbon footprint of turbines and other renewables. It's like being in an unending horror movie.

Studies were done to calculate the energy consumption and CO_2e of each mining vehicle and piece of equipment. They tracked the hours a vehicle ran and working machinery, maintenance hours, and replacement parts. Right, replacement parts; we must include the CO_2e for mining and making the replacements like tires and engine parts. And why do we need to mine for coal? Coal is burned to generate electricity for the main buildings, maintenance buildings, and whatever other buildings and processing centers a mine might have. Of course, we have to include the energy and CO_2e emissions from mining employees driving to and from work. Am I getting too picky?

Just to give you a sense of how much CO_2 there is in the mining equipment, the report estimated there were seventy thousand tons of CO_2 emissions per year (only equipment related). That is for one mine. Think about all the mines around the world. But are mines trying to do better? Many mines are aware of the environmental impact of CO_2e emissions and are starting to make small changes for the better. Just not fast enough.

Now think about all the other parts that go into a wind turbine. I don't think I can go through the CO_2e emissions for everything, but I'll give you an overview of how the carbon footprint grows with each part.

For steel:

- You need to build a steel plant,
- There is a lot of *equipment* needed to build a steel plant.
- You need a furnace to make pig iron from iron ore.
- Bring in materials that possibly use a railroad and large trucks.
- Lots of fuel for this equipment.
- Electricity needed for buildings.
- Specially made trucks to haul turbine blades.
- And on it goes.

All this and more are going to emit tons upon tons of CO_2e, and that's before we even start manufacturing steel.[105]

Hold on. Is your head about ready to explode yet? Keep on going. Cement and concrete plants must be built. There are tons of CO_2e.

Fiberglass needs glass to be processed, so that needs factories.

The fiberglass plants for turbine blades are huge and are built just for fabricating those parts. Again, tons and tons of CO_2e. Resin must be processed somewhere. More CO_2e. Oil drilling is another process that carries a large carbon footprint. You need to build the oil derricks and then the refineries. Refineries are huge and demand enormous amounts of steel. I realize there is only a small need for oil products in turbines, but you still need all of it for the different requirements of a turbine. This all goes to the carbon footprint of a wind turbine. I'm sure science is looking at making alternatives ... that work.

Can you hear people screaming, "No, you can't do that!"? Don't get your knickers in a knot, but I just did that! Now that we have our wind turbine farms, are you satisfied? Hold on to your hat! The big wind is gonna blow. At the end of its life cycle, about eighteen to twenty years, turbines get to be decommissioned, taken apart, and hauled away. More CO_2e. And guess what? You get to start all over again building new turbines. Ahh, more CO_2e. Even with better turbines, you have to rebuild them about every twenty years. Somebody's making money. The gods are telling us that turbines would be net-zero in about seven months. Does that make sense to you? I don't think they are including all the things I just talked about. So that's about it for wind, folks. For now.

Solar

Here is the next lie. Now that all the GNDers are chasing Chicken Little around trying to run him off the cliff while screaming his head off that the sky is falling, the rest of us are left to reel in these lemming politicians (GNDers) and pound some sense into them. Hello! Lemmings, listen up!

We'll start simple. What are solar cells? Also called photovoltaic cells (PV) that generate electricity from sunlight. That sounds really great. These PV cells are arranged in a frame that we call a solar panel. So, what are solar panels made of? Since there are different types of solar panels, not every panel is going to have everything on our list below, but everything on the list will be in one of the different types of panels. It's like saying all foxes are mammals, but not all mammals are foxes. Let's make that list.

1. Silicon (is in every solar panel)
2. Phosphorous
3. Boron
4. Cadmium telluride
5. Copper
6. Indium
7. Gallium
8. Selenide
9. Tin-coated copper
10. Zinc
11. Sulphur[106]
12. Silver
13. Silicon rubber or ethylene vinyl acetate to seal
14. Mylar or Tedlar backing
15. Glass or plastic cover
16. Aluminum—frame

Silicon. I feel the need to describe how silicon is made. Some people like to say it's made from sand. Sounds good, but we don't make it from sand. It's too impure. We make silicon from silicon dioxide, commonly known as quartz. It's placed in an arc furnace using, wait for it, coal. The results are silicon and CO_2. It is then treated with phosphorus or boron to make a good semiconductor.[107]

Phosphorus. It is mined as phosphate rock, also called phosphorite. It is obtained by large-scale surface mining. You know what that means, right? Drilling, blasting, removal using a dragline, electric shovels, bulldozers, and more—and CO_2e goes out.[108]

Boron. This is mined through boron-containing minerals called borates, and there are only certain places it can be mined economically. More mining.[109]

Cadmium telluride. These are two separate minerals that are sandwiched together. Cadmium is made as a by-product of mining and refining ores like zinc, lead, and copper. Telluride is a mineral associated with gold-telluride or silver-telluride. These are mined by both underground and surface mining. Then, of course, there is the processing by floatation, leaching, and other methods of separating the two. We not only have mining but the processing that is not very environmentally friendly.

Copper. We have already discussed copper in the previous section on wind turbines.

Indium. It is usually found as a trace element with other minerals like zinc, lead, copper, and tin. Indium, therefore, must be extracted from these metals. It is not in abundance, so solar panels are limited by how much indium the planet has.[110]

Gallium. No gallium is mined for the mineral itself. It is found as trace elements in other ores like bauxite, sphalerite, and coal. It is treated as an impurity and is a by-product of mining and processing. Again with the mining.[111]

Selenide. Selenide is a chemical compound. It is a compound that contains a selenide anion. For solar use, it is the compound copper indium gallium selenide (CIGS). It is a semiconductor for photovoltaic cells (PV). It is used to convert sun power to electricity. What they do is lay down a thin layer of copper, indium, gallium, and selenium on glass attached to electrodes to collect the electricity. A lot of work goes into making these CIGS cells.[112]

Tin-coated copper. Tin mineral (cassiterite) is mined. The tin is extracted by roasting with carbon (should I say ... coal) in a furnace. Then it is leached with acid solution (removes impurities).[113] Tinned copper resists corrosion and lasts much longer. They either dip the copper in molten metal or use an electric current to bind the tin onto the copper.

Zinc. Here is another mining process. Zinc is mined as an ore, often included with lead and silver. In some mines, it is a by-product. Mined ore is processed to produce concentrates, then sent to a smelter to get the zinc metal.[114] Here we go again. Mining and processing. CO_2e ... CO_2e ... CO_2e.

Sulfur. This has been mined for a long time. Early mining consisted of picking up sulfur rocks right off the ground. Starting in the early part of the twentieth century, sulfur mining became a by-product of natural gas and oil refining.[115] Researchers are planning to use a sulfur-based solar storage system, something that can be used at night while the sun is sleeping.[116]

Silver. We know how we get silver, right? Yes, *mining*, then processing. Silver is mined as an ore and then extracted from said ore by smelting or chemical leaching. Ugh! Sounds bad for the environment.[117] Imagine sunlight shining on a solar cell, making baby electrons. Silver gathers up all these little electrons and turns them into an electric current. And then silver helps shoo this electricity out of the cell so it can be used.

Silicon rubber or ethylene vinyl acetate. Silicon rubber contains silicon with carbon, hydrogen, and oxygen. We already know how silicon is made. We discussed this earlier.[118] In the solar cell, it is used as a sealant. Ethylene is a hydrocarbon (C_2H_4).[119] Are hydrocarbons bad? Of course they are. They contribute to global warming. Now, vinyl acetate is made from ethylene from a reaction with oxygen and acetic acid. I just want to tell you that from various other processes, it becomes Elmer's glue.[120]

Mylar or Tedlar backing. Mylar is a polyester film made by stretching polyethylene terephthalate (PET). It is used for a bunch of other things as well. It is derived from crude oil and natural gas. Humans use this for bottles, food packaging, and more. For the solar industry, it's used as the backing for the solar panel. Tedlar is cast using the solvent dimethylacetamide (DMAC), a highly toxic solvent.[121, 122]

Glass or plastic cover. We've discussed glass production and plastics under wind turbines. Fossil fuels. Need I say more?

Aluminum frame. Aluminum makes up only 8 percent of the planet's soil and rocks. It is mined as aluminum ore, and the best is bauxite. To make pure aluminum, you need caustic sodium hydroxide.

There are two processes to get to aluminum. The Bayer process is the first. You need a processing plant to crush the ore with caustic soda. Then it is put through a grinder and made into a slurry. Put it into a pressure cooker with more caustic soda. It comes out as a sodium aluminate solution. After a bunch more processes, it becomes alumina. You get crystals that then go into a kiln. Then it's ready for the Hall-Heroult process. I think I mentioned this earlier, but it's a smelting process. At the end of that, 99.8 percent pure aluminum is poured off. And there you have it.[123]

Is your head exploding now? The gods of renewable energy constantly remind us how great our current renewable energy is. But we cannot have this renewable energy without *fossil fuels* and extensive environmental destruction! I agree that humans will emit a little less CO_2e in the long run but not as much as the gods are telling us. Don't try to pee on my leg and tell me it's raining.

Let's get back to solar panels. There are two types of solar I'll be talking about—the regular solar panels that we see everywhere and the *concentrated solar power* (CSP). The first is the regular solar panels. They are separated as residential and commercial.

It's time we start discussing solar panel radiation. What kind of radiation? Solar panels emit electromagnetic radiation otherwise known as EMFs. But not very much. So, there are three areas of concern. The least is the solar panel itself. It emits very little radiation, and it's nowhere near our bodies. That's a good thing.

The electricity needs to go through a wiring system called an *inverter*. It converts unusable DC energy to our AC system. The majority of homes and businesses have had a smart meter installed. It monitors the amount of energy produced. Smart meters emit a large amount of RF radiation (radiofrequency) when they send data to the utility company.

Now, we come to dirty electricity. That is the unusable electricity that gets stuck in your electrical wiring system, radiating EMFs into your home. Personally, I opted out of putting a smart meter in because I don't have solar panels, and I don't want the utility companies controlling my heating and cooling system. Having said all this, there are products for sale that are supposed to help guard you from radiation: smart meter shields, protection paint, and EMF-blocking frame liner. I have no idea if they work or not.

Finally, solar panels are probably not the worst emitter of EMFs. Can it cause cancer? Probably not. However, exposure over years might increase your chances. Everyone is different. Some people have a sensitivity that causes them to get headaches, nausea, and dizzy spells. Not everyone feels this. It's a complicated issue, but people need to know what they are putting on their roofs before they decide that solar panels are the bee's knees of renewable energy.[124]

Concentrated Solar Power

The other type of solar system is concentrated solar power (CSP). The short description is a tall tower in the center of a whole lot of mirrors (thousands). The mirrors around the towers reflect the sunlight up to the tower. This concentrated sunlight boils water, kind of like a steam engine, generating steam that runs turbines that power a generator.[125] Here you have "power towers" looming up from the desert floor to heights of 450 to 540 feet like lighthouse beacons, only these towers do not protect anything.

Just a few facts. The towers contain 2,200-ton boilers. The proponents of these projects swore these facilities would not harm (or at least minimally) the environment. One facility, Ivanpah, built at a cost of $2.2 billion, is on public lands in the Mojave Desert. It's three towers and 347,000 mirrors and 3,500 acres (5 mi^2) are next to the Mojave National Preserve, the Mesquite Wilderness, and the Stateline Wilderness between the California and Nevada border. It's right in, or at least next to, the endangered desert tortoise habitat. The other problem with these facilities is the fact that the mirrors send so much concentrated sunlight up to the towers, it literally cooks any birds (in midflight) flying through. Not just one bird here and there but entire flocks that happen to fly through the area.[126]

The last fact is some CSP facilities must use natural gas to start up in the morning. At night, with no sun, the towers cool down. Some towers have no energy storage, so they need natural gas to start up fast in the morning. A few companies use molten salt to keep boilers warm. Ivanpah emits around 50,700 tons of CO_2, forcing it to participate in the cap-and-trade program to reduce carbon emissions.[126]

Just as a reminder, they are built with steel and concrete. These facilities also must have access to water for cooling. They are in the desert. All that water use will endanger all life in the desert. What happens when all the available water is used up? Or at least the water table gets too low. There are a several of these facilities in the Southwest US. Each one must clear the land of foliage, like one-hundred- to five-hundred-year-old Joshua trees and thousands of native flora, plus endangered, protected tortoises and owls have to be removed.

To my knowledge, not all these facilities are still in operation. I'm not sure, but I think Crescent Dunes in Nevada may have stopped because it lost its only income-generating US contract. Apparently, in 2019, it only delivered 0.3 percent capacity in the second quarter. Ouch! That must have hurt. Crescent Dunes was also closed for eight months because of a leak in one of its storage tanks.

In a Bloomberg article (www.bloomberg.com/news/articles/2020-01-06), a $1 billion plant became obsolete before it ever went online.[127]

We've reached the end of our journey through our solar adventure. To recap this journey, let's quickly go through the highlights.

1. Mining. A lot, I mean a lot, of mining has to happen to make solar energy happen. What does that mean? It needs bulldozers, trucks, and all the rest that goes into mining. And lots of CO_2e! And environmental destruction.
2. Harm to wildlife. CSP uses a huge amount of land. The land is also habitat for endangered wildlife. In California, Ivanpah, which is located in the Mojave Desert, is right in the middle of endangered tortoise habitat. These tortoises have been displaced, and by that, I mean they are removed from their

burrows and placed in zoo-type facilities where they may not fare so well. This also affects the endangered burrowing owls that depend on the tortoise's burrows for shelter. We now have a domino effect.

3. Loss of habitat. One-hundred- to five-hundred-year-old Joshua trees, yucca, and all the other important protected flora of the desert are bulldozed away. This becomes unlivable for the other plants and animals that have adapted to this environment.
4. Water. Production of solar cells uses water. And the CSP facilities use quite a bit of water—water that a desert community can't afford to lose.
5. Toxic materials. Making solar cells uses "toxic chemicals such as hydrochloric acid, sulfuric acid, nitric acid, hydrogen fluoride, 1,1,1-trichloroethane and acetone."[128] If solar panels aren't disposed of properly, then there is a problem. They produce "300 times more toxic waste per unit of energy than do nuclear power plants."[128]

So, there it is. If that's what you call good ideas, you're drivin' in the wrong lane. Next!

Geothermal

I kind of like the idea of geothermal. Let's see how far we get before it starts going awry, if it does. First, wells are drilled one to two miles deep. That means steel pipes and big equipment. The plants pump steam or hot water up to the surface under high pressure. When the water reaches the surface, the pressure is lowered, which causes the water to turn to steam. Again, with a turbine. The steam turns a turbine that's connected to a generator (steel) that makes electricity. The steam goes on to a cooling tower. What happens when steam cools? It turns back to water. The water is returned to the earth, and the process starts over.

This is a wonderful technological idea, but in the middle of it lies steel and concrete. Steel is especially important to geothermal because of its durability and strength. One of the biggest problems with geothermal is the corrosive nature of dissolved CO_2, hydrogen sulfide, and ammonia that's part of the geothermal process. When exposed to the steel and other metals, they are very corrosive and extremely hard on parts. Luckily, there aren't many parts, but the plants need maintenance to keep up with demand.[129]

The life cycle of a geothermal plant is about thirty years.[130] The front end of the cycle is the construction phase. The same old stuff as in wind turbines … steel and concrete, mining, trucks, bulldozers, steel making, transportation, manufacturing, and on we go. The back end is the decommissioning of a plant. Again, the machinery, the transportation, and the recycling and CO_2e.

What's so good about geothermal? Other than construction and decommissioning, there isn't a big source of pollution while it's working. It is extremely efficient. There's a lot of savings on heating and cooling. This is moving heat that is already there, as opposed to making heat. It has fewer moving parts, which makes for less and easy maintenance. Geothermal is not dependent on the weather.

What's so bad about geothermal? It's better to build newer buildings because retrofitting means large-scale excavations. The well pumps could discharge sulfur dioxide and silica into the earth. Power plants are said to be ecofriendly, but they are dangerous in that they may cause earthquakes and they are location specific. They need to be near geothermal activity, like Yellowstone National Park. They are prone to damage from tree roots and rodents, which can be difficult to repair.[131]

In the end, the worst thing about geothermal is the massive amounts of steel and concrete and the emissions for construction and decommissioning. The next is the corrosion problem. All these gases are not so good for the steel. Lastly, they could cause earthquakes. I leave it up to you if geothermal will be one of the renewables that rises in our future.

Hydropower

I'm sure we are all familiar with hydroelectric power dams, right? At least I hope most of us are familiar with this form of renewable energy. I think some people aren't too smitten with it. I'm not quite sure why. We'll see how this section goes.

Imagine floating down a lazy river in your favorite canoe. What beautiful scenery. A deer is rustling through the brush, or maybe that was a bear. Birds are chirping in the trees slowly swaying in the breeze. That river opens into a very large lake. How nice. Floating toward the far end of the lake, you see several majestic-looking castle towers looming out of the water. Waking up from your half dreamlike state, you realize you are staring at a massive concrete monster called a dam. That's the end of that ride.

What a dam does is raise the water level of a river to create a water reservoir (a lake) to use as falling water for the dam. It controls the amount of water flowing through it, and the force of the falling water pushes against turbine blades to spin the turbine. What? It appears that innovators have a one-track mind. Whatever renewable energy there is, it seems turbine technology is always there. Now, I'm not saying hydroelectric power is bad, not at all, but can't they think of anything new? Anyway, the kinetic energy of the falling water is turned into mechanical energy. The turbines are connected to the generator that converts the mechanical energy to electric energy. The electricity is transported through massive transmission towers and lines to our homes and businesses.

Two of the most familiar dams are probably Hoover Dam between Nevada and California and the Three Gorges Dam in China. I'll describe Hoover Dam, although there are many smaller dams in the US. Hoover Dam reaches 726 feet high. If you've ever been there, looking down the dam from the top roadway is like being just a little too drunk to stand straight. It is dizzying. China's Three Gorges Dam is even bigger.

Here are a few facts to ponder. Hoover Dam required more than five million barrels of concrete. That equals 940,000 tons of concrete. To make this short, I'm just going to make a list of important materials:

Concrete	940,0000 tons
Reinforced steel	22,500 tons
Gates and valves	10,835 tons
Plate steel and outlet pipes	44,000 tons
Pipe and fittings	3,350 tons
Structural steel	9,000 tons
Miscellaneous metal work	2,650 tons [132]

Not to mention the tons of iron for steel, and copper for the electromagnets. And don't forget the tons of copper cable and steel that takes the electricity hundreds of miles away from the dam. All that sent a whole lot of bad emissions into the atmosphere. Maybe that's one reason some people, maybe environmentalists, aren't so bubbly over hydroelectric.

I haven't talked about the electromagnets yet. Electromagnets are steel or iron wound up with wire, usually copper. When electricity passes through the wire, the metal is magnetized, creating a magnetic field. OK, fine. We have electricity, but it needs to move as in a current. The electric magnets are mounted on the shaft and connected to a power supply. When the power supply is turned on, the electric magnets create a magnetic field. As the shaft rotates, coils of wire are exposed to changing magnetic fields, and an electric *current* is created. That's how current moves the electricity along. At least they aren't using neodymium—just lots of copper, I imagine. Mind you, we are still discussing hydroelectric power. All of this is pretty good so far. Let's get to the not so good problems.

- CO_2 and methane form in natural water systems like lakes created by the dam. The GHG is the result of the decomposition of all that biomass that's now under water.[133]

- Building these dams means blocking and diverting the rivers or otherwise changing the natural paths of the waterways. It also blocks the migration of fish upstream and kills fish going downstream.
- What about fish migration? A Yale study showed that fish ladders were not as effective as they first thought. The ladders help adult fish, but most juveniles can't handle these ladders. During large numbers of fish migrations, the ladders aren't able to handle the great numbers of fish.[134] Many dams around the globe don't even have ladders. Some river systems have two or three dams. By the third dam, there are no fish left to go upstream. We're talking about losing entire fish populations that are an important part of our ecosystem.[133]
- Damming rivers reduces water, sediment, and nutrient flow downstream, which can lead to habitat loss and an unhealthy water system for animals.[135]
- To build a hydroelectric dam, you need to build large roads to the site, clearing land along the way. Then there are many miles of cleared land for power lines. Lots of equipment and a major loss of trees and habitat.
- Creating a large water reservoir could destroy archaeological sites.
- Droughts can potentially affect the efficiency of the dam and therefore the price of electricity. Droughts could effectively shut down hydroelectric plants.

There you have it. I don't think the environmentalists like the upfront environmental cost of hydroelectric dams as well as the environmental issues. I'm only guessing, but that's my theory. Like it or not, we only have two more renewables to look at. It's not looking so good so far.

Ocean Energy

Ocean energy is really wave energy. Proponents of this type of energy contend that it is free, sustainable and renewable and has no waste. It will reduce our carbon footprint. The basic principle of wave energy starts with the wind blowing over the ocean. Wind creates surface waves. The wave energy is converted to electricity by a wave energy converter (WEC) device.

Before waves break on our beaches, they have Brobdingnagian amounts of energy. Right now, wave energy is not a hot topic because it is not in full view of investors. There are several models of WECs. One was developed by AquaHarmonics. Theirs is a two-part device. One part floats on the surface and goes up and down with the waves. It is connected to the static device attached to the sea floor. The surface device moves faster than the static device, and the WEC converts that motion into electricity.

That motion drives a generator that makes electricity. That electricity has to be sent to a plant for distribution. Sounds pretty simple. We do have to build the plant and the devices. You know what that means. Steel, cement, copper, trucks, bulldozers, and more.[136]

Here come the negatives:

- High cost to build installations.
- It's the ocean. Lots of corrosion from the salt; therefore, complicated and expensive repair and maintenance costs.
- Noise pollution from the devices.[137] These devices put out constant noise. The rougher the ocean, the more noise is emitted. It may impact whales and dolphins that use echo location.
- Coastal erosion. The on-shore and off-shore structures may alter the currents and waves, which could affect coastal erosion.
- Besides just the high cost of construction, the devices (yes, plural, devices) are secured to the ocean floor using pilings, concrete, and chains. It also means dredging the sea floor to

accommodate the electrical cables, ripping up the sea floor environment.
- There are environmental pollutants. Problems arise with the leaking of hydraulic fluids, lubricating oils, anticorrosion paints, and coatings into the surrounding water.
- There isn't just one device; there are many located in one area. With so many devices in a delegated area of the ocean, they could impact fishing. The owners of these plants will create protected zones around the devices where fishing and boating will be banned. Fish will use this area as shelters, so fishing would most likely increase right outside these boundaries.
- Ocean ecosystems could be impacted in several ways. The area these devices are moored to on the sea floor could affect migration and movement of marine life. A lot of the sea floor will certainly be affected by all the dredging. The mooring lines could be a big problem with entanglement, especially for whales.
- The devices, depending on where they are placed, could alter the flow of water and, therefore, impact the flow of sediment around these devices. Changing the water movement could change coastal erosion and the deposits of heavier sediments like rocks.[138]

This renewable form of generating electricity is promising, but there are still technological and building challenges to deal with. We still don't fully understand the technology, and until we do, impacts on our oceans might be more than we are willing to accept. Personally, I'd like to see a lot more research and find ways to protect our ecosystems. It's great to have all these wonderful ideas, and we need more ideas—better ideas than the ones that big money and lemming politicians are trying to shove down our throats.

Biomass (Personally, I Hate This a Lot)

What is biomass? According to the IPCC, "biomass is a primary source of food, fodder and fibre and as a renewable energy (RE) source provided about 10.2% ... of global total primary energy supply in 2008." Biomass and bioenergy encompass vast and diverse topics. I'm not going to spend a large amount of time on each one but will mention some of the important ones, so you get a good idea of what's going on.

The simplest form of biomass energy is used worldwide for heating and cooking. How does that work? In the poorest communities of the world, like in Africa, India, and many others, billions of people gather wood, dung, and other types of biomasses to burn for warmth and for cooking. The IPCC touches on this, but it isn't the most important form of bioenergy they are concerned with.

Let's get to the basics of bioenergy. It comes from organic materials called *biomass* (wood, corn stalks, leftovers from a farmer's field, dung). This biomass is used to produce fuel, but it is also used for heat and electricity. I'll start with wood. It's the most important, I believe, and very problematic. In one reference, wood bioenergy is considered to be "nondepletive." That means it will always be around. Problem number one: when wood is consumed, it produces CO_2 and other emissions. According to William Hubbard,[139] this is supposed to be a "carbon neutral" process. For every tree you cut down, plant a new one. But doggone it, wait a second. Do you see the problem? When you cut down a twenty-foot tree and replace it with a seedling, it will take fifteen to twenty years before that seedling will be carbon neutral to the tree you just cut down. Think about cutting a hundred-foot tree down. A seedling could take nearly one hundred years to recover. No carbon neutral there. Once the carbon is in the atmosphere, it's doing damage. Here's the skinny of it.

> Opponents point to the potential for wood energy to increase CO_2 levels in the short run, incurring a "carbon debt" that can only be paid off slowly, and

worry that the resulting increase in atmospheric CO_2 will worsen global warming and lead to irreversible impacts before the benefits of new growth can occur.[140]

Burning trees or wood is an environmental disaster. The biofuel proponents will tell you they only take the dead, fallen trees and underbrush. That is good for fire management, but they will be lying straight up. They also cut down living trees. Check out some of the biomass power plants in Vermont.[141] They use full-grown trees, newly cut. They also add rubber chips (terrible GHG emissions) to raise the temperature of the fire to make more steam. Some so-called bioenergy plants use any kind of wood they can get, like old painted wood and creosote-treated wood. All that puts harmful chemicals into the air. If you cut down every tree in the US, it will power the US for about ten years.[141] Some humans have no moral compass because, for them, money is the bottom line.

Other kinds of biomass, which include corncobs, corn and sugarcane stalks, and lots of different grasses, which can be collected and called renewable. Again, do you see a problem arising in the future? Growing these crops every year and collecting them for bioenergy would deplete the soil for future growing.[142] Part of farming is turning over the soil with the cornstalks and grasses, as that rejuvenates the soil with nitrogen.[143] Take it all away, and the soil eventually dies. Farmers need those biomasses to keep their fields healthy. Constant taking from the land and never giving back is not an option. You will lose the fields for growing food, unless the farmers pay expensive prices to haul in new nitrogen-rich supplements. Soil mismanagement is what happened in the Midwest in the 1930s. Farmers didn't know good farming techniques to keep their land healthy, so it turned to dust.

Just a couple of last notes. What happens when this fuel runs short? Think about that. More demand for energy will require more land use for both bioenergy and food. Transportation of all this biomass to the plant, the dozers and other equipment to collect all this biomass, spews out CO2e. Yet the pushers of this bioenergy plan still say it puts out much less CO_2e than coal, which is not exactly true. But it destroys the environment much more than coal. I think we must come up with better plans than this. I'm going to leave you with one last quote.

> Bioenergy is only a renewable and sustainable form of energy under certain conditions ... To maintain the carbon dioxide balance, biomass harvest must not exceed growth increment, and carbon dioxide emitted during production, transportation and processing must be taken into account. The conversion efficiency of the product should be considered together with its end use to limit the risk of policy failure." [144]

Remember, you are unique, just like everybody else.

6

Truth about Renewables

What does all this mean? Renewable energy as proposed by the GND, especially wind and solar, were never meant to be the saviors of temperature rise and climate change. Let's see why.

I don't even know where to start this chapter. I have a huge pile of references that I haven't even used yet, screaming for me to put them in some kind of order. And more references come in every day. My plan is to start somewhere and see where we go in understanding temperature rise and the consequences of what they call climate change. In the previous chapter, we discussed renewable energy, and frankly, I don't think they can step up to the plate. Most of them are flawed, and some are seriously flawed. But politicians and media give them a pass because of big money and the subsidies politicians give out. This chapter will go over why current renewable energy gets such an easy ride. Fasten your seat belts; we're going for a bumpy ride.

We will drive through the forest of renewables. Eventually, that will take us to the ocean of conclusions. Then we'll arrive at the last two chapters about what the heck we do about this.

What have we learned reading all this? Are we confused now and don't know where we're going? This chapter will be a review of facts, theories, and lies if there are any. Let's start our drive through the forest.

Wind Turbines

I'm really itching to get into wind turbines and the other renewables in terms of the GND. We can't seem to get away from the two stars of green energy, wind and solar. That's what we hear about all the time. They aren't the only ones. We will talk about all the renewables—that is, geothermal, hydroelectric, and biomass (BOO!). You're going to love that one. We should go down the bullet points first, then discuss some of them.

- They require government subsidies.
- Without the subsidies, most if not all wind and solar farms would collapse. It's all about the subsidies.
- Can't build wind turbines without fossil fuels.
- Need for clear-cutting forests to a setback of 492 feet.[183]
- Poor efficiency.
- High maintenance costs.
- High failure rates.
- Don't do well in extreme weather or high winds.
- New turbine blades are not recyclable. They end up being piled up somewhere or shipped to third world countries for storage.
- More of these renewables mean 500 percent expansion of mining, land use, and production of waste.[184]
- Turbines must be decommissioned and replaced every eighteen to twenty years. More building materials and more mining.
- Use of SF_6. It is 23,500 times worse than CO_2. (Five grams is equal to 129 tons of CO_2.)
- More turbines mean more environmental destruction.

I've made a list of conclusions to make it easier to understand my thinking. Hope it helps. I need to start with Conclusion 9. You'll see why.

Conclusion 9: put everything together, and wind turbines become too expensive, inefficient, and ineffective.

Conclusion 10: is the amount of environmental destruction worth it? Think of how many we will need to power the country.

Subsidies are kind of like stealing money from Peter to give it to your pal Paul. Ever wonder why they need our tax dollar to give to renewables? Subsidies. It's so we can get cheap electricity right now. Once they get rid of fossil fuel energy, they'll be able to charge us whatever they want. Without a backup plan, our lights, air-conditioning and heating will constantly be going out. Once the GNDers have succeeded in planting the seed of ridding our country of fossil fuels (*but we can't get rid of fossil fuels*), where are they going to get the fossil fuels needed to make the steel and cement needed for the renewables? I guess we'll let China do all that and import what we need. When China has a real stranglehold on us, they can begin to control us. They have succeeded to a small extent already. What if they decide not to export certain vital products to us? Like pharmaceuticals. That will leave us in a great big pickle.

Don Surber of the *Charleston Daily Mail* wrote, "It is all about the tax subsidies." No subsidies, and the renewables stop running, or costs go way up for us. Much, much more than you are paying now. It is about playing on our fears, scaring us over global warming and climate change for corporate profits.[181] Socialist politicians dream of powering the country solely with renewables. I love that, too, but not with these renewables.

Conclusion 11: fulfilling this GND dream will require the greatest expansion of mining and land use we have ever seen and the greatest production of waste.

No matter how you look at it, a man wearing blinders will never see the whole picture. People, take off your blinders! Please! Look closely at these renewables. They need subsidies to keep running. What a way to run a business! I'm going to repeat myself; some people are making a lot of money with this magic flimflam show.

On one wind farm, blades fell off, and the entire farm had to be idled. The blades don't just fall straight to the ground; they fly and land anywhere. One landed in a cornfield. I've seen pictures with blades sticking out of barns. I don't know, but it might have been a doctored photo. Blades can be larger than two hundred feet long and weigh more than 18,000 pounds.[185] Then there's the work to repair those turbines or replace them. I haven't heard of humans being killed by a failed wind turbine, but the more we put up, the more likely it's going to happen. I have heard of wind farms closing. Owners just walking away. Who's going to clean up the mess or take over?

For truth's sake, we need to mention some research in different types of turbines. One is the *vertiwind* with a capacity of 2 MW. It's kind of funny looking. However, it is less efficient than the regular vertical-axis turbines. Keep trying.[186]

Conclusion 12: turbines need replacing every eighteen to twenty-five years, which means more steel, cement, and fossil fuels.

The other thing I don't like is that owners sometimes have to clear-cut forests to build wind farms. If they want to build a wind farm on hilltops, which has good wind, they will clear-cut however much land they need for the turbines, plus the 492 feet around the farm.

Conclusion 13: more clear-cutting of our forests.

The next issue is what to do about old turbine blades. Most are *not recyclable* at this point in time. We're working on it, but we're not quite there yet. So, they pile up in huge trash heaps until the problem can be solved. And then the problem that the entire turbine has to be rebuilt approximately every eighteen to twenty years. And that is being generous. The GNDers say twenty-to twenty-five years, but the turbines themselves are telling us their life span. We'll have to tear the whole thing down and rebuild it from the ground up.

Conclusion 14: Newer turbine blades are not recyclable. What do we do with them? I've heard that a company is working on recyclable blades, which is a good thing.

What about this SF_6? It's 23,500 times worse than CO_2 emissions. Just to remind you, five kilograms of SF_6 is equal to 129 tons of CO_2.

Don't forget the ocean and lake turbines. You have to rip up the bottom ecosystem for ocean turbines. Do I need to go on?

Conclusion 15: *SF^6 is 23,500 times worse than CO^2.*

Conclusion 16: *turbines are not very efficient.*

We're done with Turbines for now. I think I'm starting to beat a dead donkey here.

I'll try to go through each renewable as concisely as possible. They include geothermal, solar, hydroelectric, wind, and biomass. Hopefully, we will see what they produce or should produce. Remember, these are not 100% efficient. And we're off.

Geothermal

Most of the geothermal information is in chapter 4. So, this will be all bullet points:

- Great technological idea.
- Life cycle is about thirty years.
- No big source of pollution while it's working.
- It is very efficient.
- It's not dependent on the weather.
- It uses tons and tons of steel and concrete.
- There is the corrosive nature of dissolved CO_2, hydrogen sulfide, and ammonia that is part of the process. This means a lot more maintenance is required.
- In building and decommissioning, there is a great deal of pollution.
- It's better to build new plants than to try to retrofit older ones.
- Discharge of sulfur dioxide and silica into the earth.
- Location specific.
- Corrosive gases are not so good for steel or other metals.
- Possibly could cause earthquakes.

Conclusion 17: I leave it up to you, the reader, to draw your own conclusions on geothermal.

Solar

What about solar panels? Well, what about them? They seem like they should last forever. No big moving parts, and they only have to absorb the sun's energy and convert it to electricity. There's always a but in everything, though. Solar panels do start degrading after a while. Didn't know that did you? They are supposed to last about twenty-five years. Why twenty-five years? Because most solar companies only guarantee them for that long.

- Panels begin to degrade over time.
- Life cycle is about twenty-five years.

Conclusion 18: solar panels begin to degrade over time.
They are ultimately less able to convert the sun's energy to electricity. What's the problem? Solar panels become less efficient "due to factors such as hotter weather and the natural reduction in chemical potency within the panel."[187] If you have solar panels, keep an eye on your bills. When the costs start creeping up, it might be a sign your solar panels are starting to degrade. Can you imagine replacing all the panels in solar farms and atop houses and commercial buildings? Yikes. What do we do with all those degraded solar panels?

What about solar panel degradation? I didn't know this, but I learned there are companies that do solar risk assessment because solar projects have been "chronically underperforming."[188] It appears the discussion is about solar assets and how not to lose investors. It also appears that solar panels are degrading faster than they anticipated. Oops! They had been saying that the degradation rate was 0.5 percent yearly. It's more like 1 percent.

Then you have the big wildfires that happened in 2020 and 2021. There were the extreme cold events. They all affected solar performance. One of the reports says during the twenty-year life span, the degradation could have been underestimated by around 14 percent. Why the big deal? This performance overestimation affects revenues, especially for the investors. The result of all this, the industry's summary, was that they had "significant work to do." That, my friend, could affect their credibility as an entire industry. So, what do they do? They tell their investors to hang in there ... we're working on it. And that's that.[188]

Parts of the panel are recyclable.

- Many panels go to third world countries to be recycled.
- There is some toxicity in the recycling process.

Conclusion 19: many solar panels go to third world countries to be recycled due to the toxicity of the process.

Conclusion 20: What happens to the parts of a panel that can't be recycled? They're dumped in a trash heap somewhere.

Weather changes also have negative effects on panels because of the expansion and contraction of parts. The good news is you can extend the life of your panel by working with a good company and keeping snow and debris off the panels.

Conclusion 21: extreme weather is not so good for solar panels.

There are two main things that take up space in a solar farm—the solar panels and the structural part. The power output of a solar farm is difficult to calculate because there is a range of energy absorption depending on where the sun is during the day and the weather. Solar Star in California has an 18 percent efficiency, but the average is around 20 percent.[189] Please remember there are a lot of different types of solar farms out there. Some will have better and some worse efficiency.

Conclusion 22: You need six to eight acres for 1 MW capacity.[190]

Conclusion 23: Poor efficiency.

One MW of energy can take up six to eight acres of land. It's really difficult to determine the exact amount of energy a solar panel generates because there are so many variables, like how much sun they are getting, good or bad locations, weather, and so on. To make it more complicated, the solar panels generate DC (direct current), but we need AC (alternating current) for our homes and businesses.

Solar farms need to have inverters that convert DC to AC. The problem is that about 20 percent of energy is lost in the conversion. That goes for residential solar panels as well. A 100 MWdc current will only generate around 80 MWac. On top of that, solar panels are only about 20 percent efficient. Where does that leave the actual output of solar panels?

Does that mean that 80 MWdc only generate 16 MWac? I don't know how that is calculated. I know how to calculate how many panels an average home needs, so maybe I will work with that. According to the EIA (US Energy Information Administration), the average US home consumes about 11,000 kWh per year. The average panel is 320 watts.

There is something called a production ratio, which is calculated by dividing the number of kWh/year a system produces by the kW of the system. The number of panels the average home needs will be: 11,000 kWh/year divided by the production ratio (let's say it's 1.4): 11,000 ÷ 1.4 = 7.9 kW or 7,900 watts. The average panel being 320 watts, 7,900 ÷ 320 = about twenty-five panels. That's why the industry produces bigger and larger wattage panels.

What about solar farms? They come in many types and sizes. There are utility-size and community-size farms. Location is a big factor. Let's look at Solar Star in Kern and Los Angeles counties.[191] Its installed capacity of 579 MW of energy can power 250,000 homes. That's all I got. If its installed capacity, it must not include the 20 percent efficiency rating. Solar Star has about 1.7 million solar panels over the space of five acres. The point is how many more homes could this facility power at 50 or 80 percent efficiency? Efficiency is what we need to work on.

But right now, solar panels at best are only 20 percent efficient.

Conclusion 24: With better efficiency, we could power many more homes and businesses.

I came across this little gem recently. A Bomen 100 MW solar farm in New South Wales, Australia, had a huge disaster. They had a mechanical malfunction that caused a three-week power outage in January 2021. Issues like this are starting to make contractors, who collapsed after this fiasco, leave this business and go back to building roads.

It appears solar is all hat and no cattle.[225]

Concentrated Solar Power (Ugh!)

What next? How about concentrated solar power or "power towers"? Or as I call them, Terror Towers. Let's just cut to the chase, shall we? CSP has a crapload of problems. There are water issues. So, what do they do? They build these humongous things in the desert where water is at a premium.

- They use so much water that it can lower the water table.
  Conclusion 25: plants itself in the desert and uses up mega amounts of water that affects the local communities.
  Many of these towers have a laundry list of issues. Mind you, there are a lot of different styles of CSP towers, but they all have similar problems.
- Land use. The new Ivanpah Terror Towers occupy five square miles of Southern California desert. There are 347,000 mirrors the size of garage doors tracking the sun to focus the heat energy on boilers in three 450-foot towers.[193] With all that, it's not producing enough electricity to fulfill its contract. They're always running the risk of being shut down.
  Conclusion 26: they take up a huge land mass for what they do.
- They aren't producing enough power to fulfill their contract. Their efficiency is between 7 and 25 percent.
  Conclusion 27: they aren't producing anywhere near the requirements due to low efficiency.

- Five square miles of land have to be leveled, so the five-hundred-year-old Joshua trees and native plants have to be bulldozed out.

Conclusion 28: terrible environmental destruction.

- They told us there would be no damage to the environment.
- Use of hazardous materials. Generally, they used water to clean the mirrors and for cooling. Since water is at a premium, they're changing to other cleaning agents, which include "hydrochloric acid, sulfuric acid. nitric acid, hydrogen fluoride, 1,1,1-trichloroethane, acetone, and others."[194] All of which probably drip into the soil. In their defense, they say they are looking for cleaner avenues for the mirrors. Who knows? Remember, they are also getting a lot of money from the government to keep it going.

Conclusion 29: use of hazardous materials that could leach into the ground.

- Ivanpah is in the Mojave National Preserve and other wildlife preserves. Federal wildlife officials have said these towers were "mega traps" for birds and other raptors. The other problem for these facilities is that they are located in the endangered tortoise area. Tortoises live in variable environments but are drawn to water and will travel a long way in search of food. Their greatest threats are human activity. These towers certainly provide that.[195] And now the best for last. There are small owls that live in tortoise burrows, and when the tortoises are removed and relocated, these owls lose their homes.

Conclusion 30: traps for birds (by the flock), built in endangered tortoise preserves, which also affect the ground-dwelling owl.

- When the sun goes down, the mirrors no longer function. The boilers cool down at night. Something is needed to keep the boilers warm, or it takes too long to get started in the morning. And that, my friend, is done by burning natural gas. For these terror towers, 63 percent of the energy is from

solar, and 37 percent is from natural gas. I'll bet the elite government officials never told you that. I'll bet they don't know. Remember, we are getting rid of fossil fuels ... except maybe for what they are not telling us about.
Conclusion 31: mirrors don't work at night.

- Another CSP facility is Crescent Dunes in Nevada, USA. Now there's a story. Solar Reserve developed Crescent Dunes CSP. I believe it has stopped operations over a power-purchase agreement. NV Energy had agreed to buy electricity from Crescent Dunes until 2040 but cancelled the contract because of performance failures.[196] In less than a year, repairs to fix a leak led to an eight-month shutdown. Apparently, prolonged outages reduced the expected energy production by 50 percent in 2019, and it ended in bankruptcy.[197] So, that's that. I think it was bought up by some other company.

There are facilities that use hot molten salt to keep the boiler warm at night so the tower can get a good start in the morning. A good cup of molten salt really gets me going in the morning. But I'll bet you they still need natural gas as a backup in case of an accident with this molten salt. Just my opinion.

There are energy storage abilities for CSP facilities. The different options have their own set of problems. Here are six options and their problems:

- Solid particle.[198] The particle-to-fluid heat transfer can be very problematic.
- Molten nitrate salt. The problem here is it decomposes at about 600°C (1112°F).
- Batteries. Very expensive.
- Pumped hydro. You need very large amounts of water. Bad for the desert community.

- Air compressed. Unique location is required.
- Flywheels. This only provides minutes of storage. Sometimes only seconds.[199]

Conclusion 32: it appears that energy storage for CSP has its problems.

People get set in their belief that these renewables are the answer to our climate problems. But they aren't. Like a tree trunk that hardens, it becomes impossible to bend and eventually will break in the wind.

Biomass (Another Ugh!)

Why in the world this is called a renewable, I'll never know. Whoever is convincing the powers that be that biomass is a good way to go, probably powerful lobbyists, is not working in the best interest of climate change or the people. Politicians who allow this technology have got to be getting something in return. "Common sense is like a deodorant; the people who need it the most never use it" (anonymous).

I'll get right to it. The claim is that biomass is low carbon or carbon neutral. That biomass doesn't contribute to climate change. Yes, it does! Burning biomass emits more CO_2e than fossil fuels per MW of energy produced.

Natural gas emits 117.8 pounds CO_2/mmbtu (1 million btu) (0.3MWh).

Coal emits 205 pounds CO_2/0.3 mwh.

Wood emits 213 pounds CO_2/0.3 mwh.[200]

Conclusion 33: biomass emits more CO^2 than both coal and natural gas.

These facts are not controversial. Biomass air permit numbers bear this out. Biomass plants are required to release their GHG emissions, I believe to the EPA. Here's another handy dandy set of facts. Let's talk boilers.

Biomass boiler is 24 percent efficient.

US coal is 33 percent efficient.

Natural gas plants are 43 percent efficient.

According to EPA rules, a biomass plant cannot exceed 1.6 tons of CO_2 per MWh per year.

Aside from the facts I just gave you, proponents of biomass energy will give you two arguments: the waste argument and resequestration argument.[200] I'll try to make these as short as possible.

- The waste argument, in a thimble, says, "it (wood) would have decomposed anyway." Why is that wrong? Because of *time*. Biomass facilities add all the CO_2 immediately and continues whereas it takes years or decades for trees to decompose where they fall. CO_2 is slowly released over time, which is the natural order of things. Some of that CO_2 will return to the soil as well. So how can biomass energy that immediately releases all that CO_2 into the air be carbon neutral? It can't. Time makes all the difference.[200]

Conclusion 34: With natural decomposition, CO^2 is released over time. With biomass burning, CO^2 is released all at once.

With a growing biomass industry, will logging and milling be able to supply all these biomass plants, as well as waste facilities and construction waste? Come to find out, a lot of facilities are already cutting down trees for fuel. This means clear-cutting. They use painted wood and creosote-treated wood. They release other nasty gases as well.

Conclusion 35: If we can't feed these biomass plants with logging waste and other waste, they will clear-cut forests for their fuel.

Conclusion 36: Biomass facilities use painted wood and the toxic creosote-treated wood.

There is a part two of the waste argument.

- The production of methane. It's called the "Methane Myth."[200] They will tell you that it's better to gather the logging materials for biomass fuel because letting this stuff decompose the way nature intended can emit methane. Number one, methane is produced in environments like wetlands or swamps. It's not produced in the logging areas where the methane emissions are very small.

Landfills are a source of methane. There is a study that shows "the resistance of most forest products to anaerobic decomposition is significant."[200] Only 3 percent of land-filled wood is emitted as methane and CO_2.

Conclusion 37: Methane myth—letting logging materials decompose emits methane. Large amounts of methane are only produced in swamps and wetlands.

- The last important issue is huge. What about the towering piles of wet, steamy, badly aerated mountains of wood fuel stored at the biomass plants? Gases emitted from these fuels produce "in addition to CO_2, carbon monoxide, methane, butane, ethylene, and other toxic gases." Could that be an accident waiting to happen? Not to mention the tire chips some facilities are using to stoke the fire to make it hotter. According to the Environmental Law Alliance Worldwide (ELAW.org), burning tires has serious health risks. These emissions spew out benzene, 1,3-butadiene, and benz[a] pyrene, all known carcinogens.[201] Like I said, money is the bottom line. The owners of some these facilities are not in it to save the planet or the environment. They would clear-cut every tree for miles and miles if allowed to.

Conclusion 38: huge piles of fuels produce large amounts of CO^2, methane, and other toxic gases.

Conclusion 39: using tire chips to fuel the fire releases carcinogens and poses health risks.

- The second argument is that of resequestration. The idea that biomass energy is carbon neutral because replanting trees will recapture or resequester the same amount of carbon released to the air by burning biomass fuels lacks common sense. They want us to believe biomass is net-zero carbon. The problem with this argument is again *time*. "A 50-megawatt biomass power plant burns more than a ton of wood a minute. It takes seconds to burn a tree." How long does it take that same size tree to grow back? It takes decades or more to grow a tree.
- Then they will argue that as long as there are forests growing more wood than is being cut, then CO_2 emissions are neutralized. This argument is again *wrong*! The forests that are growing right now are taking up carbon. By cutting and burning these trees now, it decreases the carbon uptake by trees and thus increases the carbon emissions from biomass energy.

Conclusion 40: planting young trees to replace the large trees does not equal carbon neutral.

- The next thought you need to mull over is a report by the Manomet Study that concluded producing a certain amount of biomass energy would emit twenty tons of carbon and producing the same amount of energy using fossil fuels would emit eleven tons of carbon.[200]

Conclusion 41: biomass produces more CO^2e than fossil fuels for the same amount of energy.

With all this going on, maybe this would be a good time to include and discuss electric vehicles and the Keystone Pipeline. It's well worth the drive now that we have all these terrific renewables.

Electric Vehicles

Just think, we could all be driving electric vehicles, and city transportation could be all electric. Just a tidbit of information here. I'm not against electric vehicles. By all means if you want one buy one. Most cities had electric public transit in the early twentieth century. They ripped it all out. We'll need charging stations for residential, public, and commercial use. Wherever we park our cars, we'll need a charging station nearby. Let's go through the requirements of the electric vehicle (EV) and stations.

As our EVs begin to hit the roads, steel will grow right along with them. Steel plays a very important role in the chassis, engines, and charging stations. Recently, steel has been used over aluminum because steel costs less and has more strength. There's something called advanced, high-strength steel. It lowers the weight yet provides better passenger protection. They have also developed electrical steel. It has certain magnetic properties that can be used in transformers and generators. Maybe for wind turbines? You got it. However, it is also used for the stators and rotors for EVs.[145] And remember, charging stations need concrete, plastics, rubber, and oodles of copper. Take a good look at one when you come across one. How does all that electricity move around? Copper.

There's a problem. They have to create a large enough charging station network. We don't have nearly enough. As more EVs hit the road, there could be enormous congestion at these stations. There are a whole bunch of problems in the installation process. Even before charging stations can be installed, here are some of the issues:

- There is the permit process. Lack of knowledgeable of inspectors and high costs.
- Lack of building codes regarding charging stations.
- Lack of knowledge about electric vehicle charging stations (EVCS). We're going to need much more training.
- Lack of understanding by the consumer about EVs and station needs.

- For multiunit buildings, the managers and owners need to learn about the difficulty associated with the installation of charging stations.
- Businesses will have to learn about installing charging programs.[146]

Now comes the dirty little secret about some of the minerals and elements associated with electric vehicles. They need cobalt and lithium for sure. They are part of what makes up the batteries for EVs. The secret comes in play when we talk about mining. Cobalt is found in the earth's crust but only in combination with other chemicals. To get the cobalt, it has to be smelted. Cobalt is usually a by-product of copper and nickel mining. It seems that over 50 percent of the cobalt comes from the Democratic Republic of the Congo. Why do I bring this up? Cobalt is mostly used in lithium-ion batteries.[147]

Apparently, there are two types of mining for cobalt. One is a government-verified mine, and the other is artisanal mining. Tens of thousands of children are drummed into labor for artisanal mining. There is little to no planning. The workers dig with hand tools, sometimes hundreds of feet down. There are few, if any, safety precautions, which result in injury and death. It pollutes the water and exposes wildlife and the communities to toxic metals. This results in fighting between the mine owners and the indigenous people. When Apple Inc. heard about this, the company decided to only buy cobalt using verified suppliers. Here's the rub: getting the cobalt from verified sources won't solve the problem. Artisanal mined cobalt somehow always gets into the supply chain.[147, 148, 149] If you feel anything, feel for these children.

So, that's cobalt. Let's go on to lithium. Driven by some mad, mad desire to transition off fossil fuels, the joke is *they* need money to invest in lithium to make the rechargeable batteries and EV batteries. What they don't know or don't care to know is the cost to the environment. Three countries produce about 70 percent of the world's lithium: Chile, Bolivia, and Argentina. They call it the Lithium Triangle.

Lithium comes at a terrible environmental cost as well. It's mined by pumping huge amounts of water through the earth's surface. This salty brine water is left in very large pools to evaporate. What's left is basically lithium. However, it takes 500,000 gallons of water to extract one ton of lithium. In some regions, dry conditions make it very hard on the indigenous people. The mines have used up around 65 percent of all available water. Here we go again. There's conflict between mining and the survival of the people and the environment.[150]

There exists the very real possibility that toxic chemicals will leak from the evaporation pools to the water supply. Chemicals like hydrochloric acid are used to process lithium. This harms the soil and causes air contamination. According to a lithium battery expert from the University of Chile, "Like any mining process, it is invasive, it scars the landscape, it destroys the water table, and it pollutes the Earth ... this isn't a green solution—it's not a solution at all."[151]

Get a load of this last comment. California couldn't meet its electrical needs with renewables, so we had huge rolling blackouts for days and weeks at a time. There goes the food in my fridge. Drat! Now, I'll need to buy a home generator. Recently, in California, EV owners were asked *not* to recharge their cars because we were in a drought crisis. They wanted everyone to buy EVs, and now we couldn't charge them. I heard that some owners have gone back to gas cars.

We haven't touched on electric vehicles very much. A car battery can weigh as much as 500 kilograms (1,100 pounds). Making them means digging up a lot more minerals. It means mining and processing millions of pounds of toxic materials on this planet. Our environmentalists will not allow this Biden administration to mine in the US even if it means the loss of many more jobs. In fact, the US has the best safety record for mining rare earth minerals and copper. The president met with the mining associations to tell them they will be doing more mining to be part of the electric car production. That gave him their vote. After his election, he announced the US would be importing the necessary mining products to make electric vehicles. Let's import these minerals from countries that have little or no environmental safety precautions and use children as miners.

What else can we do? We can use gasoline and mine one-tenth as many materials (destroying less of the environment) and deliver the same number of miles over the seven-year life span of the battery.[202]

Here come the bullet points:

- EV makers have chosen steel over aluminum for the chassis because it is stronger (to hold up that 1,100-pound battery) and costs less than aluminum.
- There needs to be many more charging stations, which means more concrete, steel, plastic, rubber, and oodles of copper.
- Need to grow a network of charging stations, which means training for installers, maintenance and repair, and training for companies, individuals, cities, and counties who will be using the charging stations. Yes, more jobs, but it will take years to train this army.
- Much more mining for copper, nickel, and lithium.
- Lithium mining has a terrible cost on the environment.

Conclusion 42: for now, most EVs are powered by coal.

Conclusion 43: EVs will still need concrete, steel, plastic, rubber, and copper for charging stations.

The last sad note is that wind and solar together won't deliver enough power to support all the EVs they want to sell us. Maybe this doesn't sound as good anymore. And I like EVs.

Keystone Pipeline

I'm going to blurt it out. What does the cancellation of the Keystone Pipeline mean? It *doesn't* mean we will stop receiving oil from Canada. The pipeline would've sent us more oil, and faster and safer. Unfortunately, the progressive politicians and environmentalists need to understand that the cancellation of this pipeline was just a sham show for their benefit. It was done to appease the environmentalists and anti-fossil fuel activists into thinking the government really cares about getting rid of fossil fuels. It's only to get votes. The government knows we can't get rid of fossil fuels. "They can give you money to make you richer. They can make you smarter and taller and get the chickweed out of your lawn" (P. J. O'Rourke).

The pipeline was very important to continuing a good relationship between Canada and the US. The pipeline would have solved a lot of problems, and it caused Canada to get really upset with us. So, what now? Does that mean we have no more oil? Or a lot less? That's what the progressive government would like you all to believe. See what we did? We're going fossil fuel-free! Not exactly. The Whitehouse is now pleading with OPEC to produce more oil. Now, we will go to crude-by-rail. That gets very complicated. Everything depends on the cost of crude oil, cost of shipping by rail, and the price of crude on the global market.[203]

Do any of you think the Keystone Pipeline is the only pipeline working between Canada and the US? Or the only pipeline in the US? If you believe that, then I dare you to go find a fart on a tornado. All the government did was put eleven thousand Americans out of a job. When they cancelled this pipeline, it only made it harder to get the oil to the refineries because of the rail systems. Pipelines run twenty-four hours a day and seven days a week. The Keystone Pipeline was being built in sections, with sections that ran all the way to Texas. What the government elites don't tell you is that even without Keystone, the US is set for record Canadian oil imports. It's called ... other pipelines and trains. Ouch! Oil traders and analysts say the US-Canada pipelines have enough capacity to bring in more volume of oil.[204, 205] But that might take a little time to get going.

The US government has apparently brokered a deal between the Turkmenistan government and the Taliban to put in a trans-Afghanistan pipeline to cross Afghanistan and Pakistan to India. Turkmenistan is known for its environmental disasters and lack of civil liberties. It's a dictatorship we are helping with a pipeline.[206] But we have to shut ours down. So, all this talk about fossil fuels and climate change is nothing more than theater, so take a huge bow, Mr. Hypocrisy.

Conclusion 44: stopping the keystone pipeline was a flimflam show that put eleven thousand people out of work.

Conclusion 45: we are helping to promote a trans-Afghanistan pipeline for a dictatorship and the Taliban.

What is the matter with this administration? Did we all know this was going on? The secret is we don't want fossil fuels, mining, and manufacturing in the US because of their CO_2e emissions. Let other countries do this dirty work, and we'll make our environmentalists happy because we will be at zero-carbon and get their vote.

Conclusion 46: The US government doesn't care about global emissions, climate change, or temperature rise. They only care that the US looks good and that their environmentalist base is happy.

Conclusion 47: based on conclusion 46, there will be a tremendous loss of jobs.

I'm trying hard not to mention any names, but like I said before, some people are making a lot of money. Ferment all this, and we get to the big conclusions in chapters 7 and 8.

Beware the bribed man.

7

What Sort of World Are We Headed For? (And Is All Hope Lost?)

With energy production and consumption, nuclear energy, renewables, coal, and carbon capture in the mix, what should we be doing to lessen our CO_2e emissions? What will population growth do to our future?

One of the most important pieces to our global warming and climate change puzzle is population growth. It is growing at a speed that rivals Superman. This is something not being discussed by the government or the media. Maybe it's taboo. We need to talk about it and see what it could do to the future of humans. Let's see how a growing population meshes with extreme weather events and where that affects humanity in the future.

Population Growth

Before we get into the rest of the chapter, I want to show you the correlation between the rise in population and the rise in GHG emissions. I found data for the population in 2020 along with the number of households and CO_2, methane, and SF_6 emissions.[177] I graphed them so you can see pretty easily how the rise in population is one of the causes of our problems. Besides the population, our demand for energy and resources keeps rising, and we humans don't seem to be aware. It seems we are only given one option ... use our current renewables. That will solve our issues. Chicken doodoo.

See appendix VI—Population Graph.

Population growth is very important because it affects the planet's ability to handle climate change. We will be less able to absorb greenhouse gases through clearing more land for agriculture to feed nearly ten billion people. How about housing? More land for expanding cities. More emissions. Where will it all end? Population growth is one of the drivers for climate change.[226]

As our population continues to rise, our ability to provide food will become more difficult, as more land will be claimed for renewables. Eventually we won't be able to provide enough food for a rapidly growing population. What follows is global famine. We will all look to our leaders, wondering why they didn't see this coming or didn't care. I guess taking gifts from lobbyists and others is more important, so very little will be done until the people fight back over what is happening. We will find out that wind and solar really aren't the answers. These renewables are about as useful as a trapdoor in a canoe.

Conclusion 2: Without a solution, the population will outgrow our ability to feed that population and even meet the energy needs. Isn't that a bummer!

Sea Level Rise and Migration

According to "Ask Ethan" on *Forbes*, since 1880, a 0.98°C[178] change in temperature has given rise to about a six- to eight-inch increase in sea levels. That doesn't sound like much, but it has an effect. Low-lying areas get flooded. Who wants to live where you get flooded all the time? I'd be moving out of there. I know. Grew up there. Family home. Great memories. I really do understand that. People have migrated since the dawn of humans. I'm sorry if I sound harsh, but that's what living creatures do on this planet Earth. Migrate.[179] My mother migrated from Belgium at the end of WWII to Louisiana, USA—after getting malaria and being blind for three weeks after swimming a flooded river with a baby, after a hurricane. (She didn't know how to swim.) With me in her arms, she then moved to California, USA. That's migrating.

Conclusion 3: Migration is something every living thing does at some point in time. Humans today do not want to migrate because they are financially bound to their land.

Increasing Wildfires

From chapter 3, we know that with all the statistics, you have to decide for yourself whether wildfires are due to fossil fuel use or just plain human carelessness. Here is the information listed in simple form:

- Nearly 85 percent of wildfires in the US are caused by human activity.
- The other 15 percent are due to natural causes like lightning.
- Human-caused fires are due to campfires, burning debris, cigarettes, equipment malfunction, downed power lines, arson, and more.
- Our growing population means more humans tramping through our forests and therefore more fires.
- Forest managers often allow fires to continue burning if not in populated areas. More CO_2 and nitrous oxide are emitted.
- Nitrous oxide is acidic, which contributes to acid rain affecting our coral reefs.
- In 2018, 765.6 million acres of timber were burned.
- One and a half billion tons of CO_2 were released due to wildfires, approximately. Take into account the poor forest management.
- Therefore, of the 58,083 fires in 2018, 85 percent were caused by human carelessness, not fossil fuels.

- There were 8,713 fires caused by nature itself.
- Chapter 3, figures 10 and 11 shows *natural-caused fires* have not increased over time. Therefore, no effect from global warming or climate change.

Conclusion 4: You can blame wildfires on human carelessness, not fossil fuels. Natural wildfires have stayed steady over time, so climate change has had no effect on natural wildfires.

Hurricanes

Back in chapter 2, we discussed hurricanes. Refer to figure 1.

- We have always had hurricanes. There will be more.
- Before the 1940s, only detailed descriptions were available.
- In 1943, airplane reconnaissance was started.
- In the 1970s, the Saffir-Simpson method for measuring strength of hurricanes was introduced.
- Referring to figure 1, the top line of the graph (1) shows a rise in the number of named storms with each new form of technology being developed. These are mild hurricanes.
- The line 2 represents category 1 and 2 hurricanes. Notice there isn't much change. Numbers go up and down but very little rise.
- Line 3 represents categories 3–5. Here, there is a very *slight* rise in numbers.
- Numbers rise as technology develops.
- The year 2005 is an outlier year—one of those freaks of nature we can't explain, but it can't be blamed on climate change because it doesn't happen every year.

Conclusion 5: fifty years more are needed to track hurricanes to see a good trend.

Conclusion 6: categories 3–5 have barely risen in the last one hundred years.

It's easy to see from figure 1 that when technology appeared, the number of storms increased. We were able to find and investigate storms better. The population went way up, so we were able to spot storms better. So, the number of storms increased. It also shows that in 2019, storms declined. I'm sure they will go up again. If we really want to get good data, we need to start our data points from when technology kicked in.

Where do we go from here? We should throw in this one article by Ellen Wald on a report by the UN. If it doesn't go in here, it will probably be forgotten. Apparently, the UN has recently released its Emissions Gap Report 2020. It is very interesting. For most countries, emissions don't look so great, even with some minor decreases in emissions. For the United States, the UN says we need *not* join the Paris Climate Accord. We should just continue what we are doing, and the UN report reflects this.[180] However, using the current renewables will not meet our goals. France and Sweden have exceeded the IPCC's emissions cap because they use nuclear energy as well.

Tornadoes

Chapter 2, figure 2 graphs tornadoes from 1953 to 2019. It appears to show an increase in tornadoes as the years go by.

- As the population increases, more sightings occur.
- In 1973, Enhanced Fujita Scale was introduced.
- The average number of tornadoes between 1973 and 2019 was 1,114.
- Stretched across the graph, there are nine points below the average line and ten points above the line. This shows there isn't a trend at this time.

- We need another fifty years of study to see a good trend.
- As the population increased and technology improved, tornadoes became more visible.

Conclusion 7: we need at least another fifty years of observations to determine a good trend in the number and severity of tornadoes.

Snowstorms and Blizzards

Let's take a look at the current extreme winter weather. All across the US and the globe, there's a huge winter freeze going on. Record snowfall (2020). OK. That must mean we don't really have a climate change. Or do we?

Not just Texas, where the power went out and the fallout from all this freeze has caused a blame game. One side says, "Look! Your renewables don't work in extreme weather." The other side says, "The fossil fuel plants went down as well because the state government didn't upgrade over the years, so it's their fault." The comeback is that the fossil fuel plants went down because there was way too much demand, because the renewables went down. Come on, folks! You are all right to some extent.

We have to keep politics out of all this and concern ourselves with helping people. When this winter is over, get your asses in gear and fix the problem of the fossil fuel plants. They must be brought up to speed. And now we know the renewables go completely wacky wonky in extreme weather. Or maybe the global leadership already knew this. Just thinkin'.

The second part of this weather issue comes in the form of the experts(?) quickly coming out and saying that the extreme winter weather doesn't go against their global warming models. What a bunch of twaddle, drivel, and clapped trap. Now I've heard everything. According to Noah Diffenbaugh, climate scientist at Stanford University, "We haven't had enough warming to eliminate cold events, and we shouldn't expect to have enough warming to eliminate cold events in the mid-latitudes for some time."[182] Wow! What did he just say? We *haven't had very much warming*, and we won't for some time. Looking into his accomplishments, he was a lead author of the IPCC. Wow! No wonder he quickly came out with statements. He mentions preindustrial temperatures. Is he talking 1750s or 1880s? He admits our temperature rise so far has been minimal (NOAA/Climate.gov). Yet other experts put the temperature rise around 1.1°C (1.98°F). I guess when you need to convince the people we are about to french fry ourselves up and that we only have nine years left (John Kerry), 1.1°C (1.98°F) sounds more dire than 0.98°C (1.76°F) which sounds worse than 1.33°F. I showed you my data; now show me yours. Don't lie.

- The 2020 Texas blizzards show renewables went dead. So, we know renewables don't work in extreme weather.
- Noah Diffenbaugh states we haven't had "enough" warming to affect our cold weather patterns.
- The number of blizzards has doubled in the past twenty years despite all the cries of global warming.
- We have had *major blizzards* since the Little Ice Age.
- Scientists are still trying to understand the ins and outs of blizzards.

Conclusion 8: we still need fifty years or more to study blizzards in terms of global warming and climate change.

We are almost at the end. With everywhere we have been, where do we go? Do we simply keep on going with the flow and get nothing accomplished? Politics is full of disingenuous people who say one thing but do another. We're going to discuss how we can lessen our CO_2e emissions outside of a lying, conniving, and un-American socialist political arena. What would I do to clean up our CO_2e emissions? I'll get to it, eventually.

What's left to talk about? Let's start with our energy production and where it might take us. That will lead us directly into how we might lower our CO_2e emissions, with our great hope in carbon capture and storage and then the big reveal.

Lower CO^2 Emissions

1. Energy Production (See Appendix V—Energy Production)

The GNDers advocate removing all fossil fuels. Either they are on drugs or they have no idea about the science of renewable energy. They hate nuclear and would very much like to phase that out. They are not fond of hydroelectric or geothermal either, but why not use them since they are already there? At least until they wear out. Then decommission them. Then there is biomass. You know what I think about that. I'd get rid of that today! If environmentalists like biomass energy, then they are fools.

The numbers presented here are from the US Energy Administration. I simply plugged their numbers into our already known formulas to achieve our answers of MW/year. If you refer to appendix V—Energy Production, we will retrieve several facts. In 2019, total energy production was 101 quads, and renewable energy was about 11.6 quads.[13]

Quads = quadrillion btus. A simple conversion brings us to 101 quads = 3,379,016 MW/year. Referring to appendix V, you can see on page 1 the total energy production (2019) for nuclear and fossil fuels.

The renewables' total energy production was 11.6 quads = 388,086 MW/year. You can now see the breakdown for renewables for 2019. This information is important because if we know the breakdown of energy today, we might get a better idea of what energy might look like in the future, especially with the GNDers wanting to be fossil fuel–free by 2050 or even 2030. Some officials think carbon-neutral would be fine, while others who are more radical want all carbon from fossil fuels gone from the US.

In appendix V, I played a little game. Let's suppose the GNDers got their wish and had almost all energy production coming from renewables. I say *almost* because most likely we will still use nuclear at that time. There are no carbon emissions with nuclear, or at least very, very little. Nuclear is presently at 8 percent of the total energy production. This means renewables would have to be around 92 percent. How do we break this down? With a sledgehammer.

The EIA (Energy Information Administration) projects the 2050 energy production will be 117 quads or 3,914,305 MW/year.[13, 208] (appendix V, page 2)

Personally, I think it will have to be higher. The EIA also projects that nuclear will decrease a small amount. You already know I hate biomass burning, so I'd like to get rid of it altogether. Sorry, you liars about biomass. All we have to do now is break everything down to percentages, then energy. That I have done for us on page 3 of appendix V. Follow along.

The beginning pages of appendix V gives statistics for 2019. We need to look at 2050. What does that look like? The breakdown for the total energy production for 2050 is on page 2 and 3 of appendix V. Just going through the wind turbine issue, if wind is 35 percent of our total energy production, then it produces 1,370,007 MW/year. That would be actual production, not capacity (page 3 of appendix V). Remember, we are still playing our game and giving the GNDers their wish ... almost.

If you all have followed this so far, let's figure out how many turbines are needed and how much land is needed. We have already learned that a 2.5 MW/year turbine only produces about 0.6 MW/year. If that's the case, and the production *is* at 1,370,007 MW/year, page 3 of appendix V calculates that we need 2,283,345 turbines. Then we calculated how much land we needed for that many turbines. It turns out to be around 570,836 mi^2 ... approximately. I've seen another calculation that comes in at 838,043 mi^2.

California has 163,696 mi^2.

Well ain't that a thousand barrels of farts at the Oscars! We could do the same for solar panels, but they have even less efficiency (around 20 percent) and would need lots and lots of land as well. All this is very depressing, don't you think?

2. Lower CO^2e Emissions

We have come a long way on this journey of discovery. Give yourself a big high five. We have traveled through bug-infested swamps to the blazing heat of the deserts and to the sun and back. We've looked at our renewables with a magnifier. Talked about electric cars. And we've glimpsed at the money trail which is coming up. Our discussions were not always pleasant or easy to hear, but we had to have them. Everyone needs to hear this. Calling all humans!

Since the socialist elites only blame fossil fuels, this would be a great time to throw a monkey wrench into the GNDers' conniving. Way before the time of fossil fuels, "humans may have triggered a massive but mysterious 'carbon bomb' lurking beneath the Earth's surface."[207] Long before the industrial era, before 1750, there was a huge conversion of peatland to agriculture. It could have spewed out more than 250 billion tons of CO_2. If you can't equate that to today's emissions, it is equal to more than the last seven years of emissions by fossil fuels.

Why the fuss over peatlands? Peatlands only occupy 3 percent of the land globally, but they store 30 percent of the soil carbon. Since this very old type of emissions was ignored, it seems to me that "we lack a complete understanding of how the Earth's land surfaces are driving the warming of the planet." We may have a problem here. The Paris Climate Agreement wants to use land to store GHGs.[207]

Peat normally takes carbon out of the air. If it is drained or disturbed, like digging it up for agriculture, which has been done since ancient times, it will spew out a massive amount of CO_2. The GNDers don't care about ancient times or even time before 1919 and the effect it had on our temperatures. Why should we care about that? It seems that when peat is disturbed, the carbon in it combines with the oxygen in the atmosphere to become CO_2. Ain't that a hoot and a holler. You know what else is a hoot? This CO_2 can "stay in the atmosphere for hundreds or even 1,000 years, meaning that peatland conversions from long ago ... can still be affecting the planet."[207]

It seems every time we bring up any subject, everything sounds worse and worse. On the flip side, besides renewable energy, what can we do to lessen CO_2e? There is a lot of research going on in many areas of science. There has been for many years. We just have to get politics out of it.

The easiest thing to do is plant more trees. We all know that planting billions of trees isn't the entire answer, partly because with current renewables, we are cutting down a lot of trees. If you are one of those who really wants to address climate change, look for some of these new technologies and help fund that research so we can get them going.

Carbon Capture

Now is the time to talk seriously about carbon capture or what's better known as carbon capture, utilization, and storage (CCUS). So, what is it? CCUS is the process of removing carbon from the air or from factories, mining, and anywhere that CO_2 is released and using it or storing it. Problem? These technologies are awfully expensive; therefore, there is no incentive for industry to install carbon capture systems. Until the cost of releasing carbon into the atmosphere goes up, no one will want to install it. If the government was serious about climate change, it would work on this aspect of carbon capture. That means levy high fines on these industries. But, again, lobbyists would nix it with the politicians.

Carbon capture, utilization, and storage (CCUS) is one of those research projects. Before we get into some of that research, let's talk about the naysayers of this idea. In 2019, Simon Lewis, a professor of global change science, wrote an opinion piece for the *Guardian*. In a nutshell, he thinks carbon capture isn't going to work. His opinion is, by putting carbon capture into mining and factories, it only gives facilities a "license to keep emitting and clean up the mess later with new technologies." Apparently, politicians love this because they can yap about a 1.5°C (2.7°F) target and tell us they have plans to meet that and even exceed it, hoping all the while those temperatures will be controlled *later* on in the century. That seems to be their plan.[209] However, if you limit (by law) the amount of CO_2e these facilities can release, they may start to comply, especially if they are fined for going over their limits. Just my humble opinion. We the consumers will feel the brunt of that with higher prices, but what do you want? A clean environment or cheaper prices. Buy China, goods are cheap, their environment is a disaster.

Basically, there are two types of CCUS systems, one that prevents CO_2 emissions from leaving a facility and one that grabs it right out of the air. The earliest system was part of the utilization process. CO_2 was pumped into oil fields, and that, in turn, helped get more oil from the ground. I'm sure the GNDers spit on that idea. Another example in the US was a food-processing plant that started a project to remove carbon from their plant. Apparently, it can eventually take 1.1 million tons of carbon per year out of their factory emissions. They store the carbon underground. That sounds good to me. Doesn't it? Can you imagine millions and millions of facilities doing this? The government needs to look at this facility to see how they are doing it. Don't you think? The only downside is storing it underground.

Both these ideas should be playing a bigger role in removing carbon from our atmosphere and getting to net-zero. So why aren't these processes and other technologies like this being pushed to be used everywhere? It's a matter of economics. It's cheaper for facilities to let carbon go up the smokestack.[212] When they say put a price on carbon, are they talking about a carbon tax? My eyeballs are floating around my brain. Does that mean all of us will be taxed as well? If the price of carbon goes up, then maybe there would be an incentive to install CCUS systems. Nobody wants a carbon tax. I don't. I really don't. Politicians talk about it all the time. They never talk about carbon capture. Every once in a blue moon, I hear a politician mention it, but then it goes silent. I'm sure big-money lobbyists go to work to make sure the tax doesn't happen; therefore, the CCUS systems don't happen either. The only interest is in the GND's renewable projects, especially wind and solar. They could just tax industry and leave us lowly people out of being taxed. That's my idea. Then the price of gas goes sky high.

While all this political nonsense is going on, there is a ton of research being done on carbon capture. Some of the research caught my eye, now that the crap has stopped rolling around in my head. I'll try to keep my explanations simple.

Nature has been turning carbon into rocks for millions of years. It just takes millions of years to do it. What if we could speed up the process? A company called Carbfix captures and dissolves CO_2 in water, then injects it into the ground, where it turns into stone in less than two years. It's "cheap, economical and environmentally friendly." So says the CEO. It's kind of like Mother Nature, but instead of taking millions of years to do it, humans aren't taking that long. Carbfix is planning to "reach one billion metric tons of permanently stored Carbon in 2030."[213] I love this idea. Can you imagine if the world could use this process to capture and store CO_2?

The second idea comes from Canada. It's called Modular Carbon Capture and Storage (MCCS). The first installation will enter service at Glacier Gas Plant in Alberta, Canada, by 2023. MCCS is projected to recover about 90 percent of carbon emissions or 136,000 tons of CO_2e/year.[214] The advantage of this technology is it can be retrofitted to facilities or used as freestanding units. "Most of us walk around like everything is fine, but deep down, inside our shoes, our socks are sliding off" (anonymous).

There is another process I like. MIT engineers have created a very different way to remove carbon dioxide from the air. It uses a weird kind of battery that can take in CO_2 while it's charging and release CO_2 when the battery is discharging. The CO_2 can be compressed and sent underground for storage. It apparently costs less than other methods, and if it can be developed commercially, it might just be a technology we can use.

We've discussed a bunch of CCUS projects in previous chapters that appear to be going well. Some are doing very well. We need to start using some of these on a larger scale. Like somehow forcing

CO_2-spewing facilities to install them. The problem is these facilities don't want to install them, so their lobbyists go to work influencing politicians to agree to not put them in, so carbon capture doesn't get installed. Plus, the cost will eventually be passed to us, the people. The few politicians who can't be influenced are in the minority. Someone needs to start yelling about CCUS.

Hey! Carbon capture!

Hey! Humans!

There is an article by Richard Black, the former director of the nonprofit Energy & Climate Intelligence Unit. He proposes these scenarios:

1. "Make all energy more efficient." Finally, someone brought up efficiency! Remember, in the introduction I said remember the word *efficiency*? Well, here it is. A big yes from me. All researchers, scientists, and engineers need to start working on the efficiencies of renewables we have. They are pretty bad.
2. Make electricity production completely zero-carbon. Duh! We all would like that. It's pretty vague. That's like not explaining to someone why you never kick a cow chip on a hot day. I think we need to work on this one.
3. "Extend the use of electricity into areas where currently we burn fossil fuels directly ... heating and transport, but also some industries." I assume he means with renewables. Duh! That's the whole point. He was just filling up the page with that one.
4. Use customized solutions for industries where electrification isn't possible. One of his solutions is using hydrogen instead of coking coal in steelmaking. I came across some info that might change his thinking. Hydrogen use for high-strength steel can cause what's called hydrogen embrittlement. This is a "significant permanent loss of strength that can occur in some steels when hydrogen atoms are present in the steel and stress is applied."[210] Well, golly gee. We're off and running

like a herd of turtles. This article goes on to say that there wouldn't be enough *quality* steel to cause this problem. I'm sure China makes good quality steel for export (cough, cough) because we sure won't be making steel in the good old US anymore. His other suggestion is to use wood instead of cement for buildings. Whoever thought of this one has only one oar in the water. There go all our trees. We use cement for foundations because it doesn't rot or break down like wood. Come on, folks!

5. Invest in negative emissions to clean up what we emit. I imagine that's carbon capture and planting trees. However, planting trees will become a problem. We won't be able to plant trees fast enough.[211] And it takes years to grow. We must make an effort to get countries to plant millions and billions of trees. It's called mobilization. Also, with the massive amount of land needed for renewables, agriculture, housing, and industry, how much land will be left for forests? Let's jump on the carbon capture ideas.

We're not close to being done yet. We should relist the conclusions that are in these last few chapters and see where that leaves our renewables. Besides I'm repeating things so people can get this into their heads. Conclusions 1, 2, and 3 are at the end:

Conclusion 4: you can blame wildfires on human stupidity but not fossil fuels.

Conclusion 5: on hurricanes, we need fifty more years to see a good trend.

Conclusion 6: category 3–5 hurricanes have barely risen in the last one hundred years.

Conclusion 7: on tornadoes, we need fifty more years of observation to determine a good trend.

Conclusion 8: we need fifty more years to study blizzards in terms of climate change.

Conclusion 9: wind turbines are expensive, inefficient, and ineffective.

Conclusion 10: on wind turbines, is the environmental destruction worth it?

Conclusion 11: fulfilling the GNDer's dream will require the greatest expansion of mining and land use we have ever seen and the greatest production of waste.

Conclusion 12: turbines need replacing every eighteen to twenty-five years, which means more steel, cement, and fossil fuels.

Conclusion 13: more clear-cutting of our forests.

Conclusion 14: Most newer turbine blades are not recyclable. What do we do with the waste?

Conclusion 15: SF_6 is 23,500 times worse than CO_2.

Conclusion 16: turbines are not very efficient or effective.

Conclusion 17: I leave it up to you, the reader, to make up your own mind concerning geothermal.

Conclusion18: Solar panels degrade over time. They need to be replaced every twenty-five years or less.

Conclusion 19: many solar panels go to third world countries to be recycled due to the toxicity of the process.

Conclusion 20: What happens to the parts of solar panels that cannot be recycled? Are they dumped in a trash heap somewhere?

Conclusion 21: Extreme weather isn't good for solar panels. It lowers their life expectancy.

Conclusion 22: For solar farms, you need six acres to produce 0.2 MW/year. Not very good.

Conclusion 23: solar panels have poor efficiency, 10–20 percent at best.

Conclusion 24: with better efficiency, we could power many more homes and businesses using less land.

Conclusion 25: CSPs (concentrated solar power) plant themselves in the desert and use up way too much water.

Conclusion 26: CSPs take up huge land masses.

Conclusion 27: CSPs aren't producing near up to the requirements due to low efficiency.

Conclusion 28: CSPs cause terrible environmental destruction.

Conclusion 29: CSPs use hazardous materials that could leach into the ground.

Conclusion 30: CSPs are a trap for birds (by the flock). In California, they are built in endangered tortoise preserves, which also affects a ground-dwelling owl.

Conclusion 31: CSP mirrors don't work at night.

Conclusion 32: it appears that energy storage doesn't work so well for CSPs.

Conclusion 33: biomass emits more CO_2 than both coal and natural gas.

Conclusion 34: With natural decomposition, CO_2 is released over time. With biomass burning, CO_2 is released all at once which increases CO_2 emissions.

Conclusion 35: if we can't feed these biomass plants with logging waste and other waste, they will clear-cut forests for their fuel.

Conclusion 36: some biomass facilities use painted wood and toxic creosote-treated wood.

Conclusion 37: Methane myth—letting logging materials decompose naturally emits methane. Not exactly true. Methane is released in swamps and wetlands.

Conclusion 38: huge piles of fuel sitting at biomass facilities produce large amounts of CO_2, methane, and other toxic gases.

Conclusion 39: using tire chips for biomass fuel releases carcinogens and poses health risks.

Conclusion 40: planting young trees to replace the large trees does not equal carbon neutral.

Conclusion 41: biomass produces more CO_2e than fossil fuels for the same amount of energy.

Conclusion 42: for now, most electric vehicles are powered by coal.

Conclusion 43: EV charging stations will still need concrete, steel, plastic, rubber, and lots of copper.

Conclusion 44: stopping the Keystone Pipeline was a flimflam show that put eleven thousand people out of work just to keep the environmental base happy.

Conclusion 45: the US is helping to promote the trans-Afghanistan pipeline for a dictatorship and the Taliban.

Conclusion 46: The current US government (2022) doesn't care about global emissions, climate change, or temperature rise. They only care that the US looks good and they keep their environmental base happy.

Conclusion 47: based on conclusion 46, there will be a tremendous loss of jobs.

Conclusion 3: migration is something all living things have to do at some point in time.

Conclusion 2: without a solution, population will outgrow our ability to feed that population and meet their energy needs.

Conclusion 1: after decades of research, "we have not delayed the apocalypse by a single day."

There you have it. What can we say other than these renewables were *never meant to be the end-all*. I'm going to reference the German dilemma over renewables. Germany is one of the richest and most high-tech countries in the world. The UN and World Bank invested billions in renewable energy in poor countries. Soon after, Germany was forced to announce that it couldn't meet its 2020 emissions reduction agreements. It was going back to mine for coal. That started an opposition to renewables, and that has grown. The German news wrote, "There is hardly a wind energy project that is not fought."[217]

The resource-intensive and land-intensive renewables just haven't succeeded in Germany. The people there are pretty smart. They can see that solar farms take up 450 times more land than nuclear facilities, and wind farms take seven hundred times more land than natural gas wells (and that's to produce the same amount of energy).[217]

The people finally saw that transitioning to renewables was doomed to failure. No matter how much money is invested in renewables, they do not want to go back to "premodern life." Renewables cannot power our growing modern civilization because they were never meant to. In the meantime, some people got very, very rich. When the subsidies run out, renewables will crash and burn, or power will become very, very expensive.

The IPPC released a climate report sometime around 2018, where they found little to no evidence that extreme weather events increased due to any global warming. Professor Roger Pielke Jr. from the University of Colorado reiterated the IPCC reports, finding that "there is little basis for claiming that drought, floods, hurricanes, tornadoes have increased, much less increased due to" GHG. I don't think you are going to hear that from the media, especially the ones that have been spouting that all our weather events are proof of man-made warming.[218]

We've already read about all the minerals we need to mine to keep our renewables and EVs industries growing. The World Bank estimates that the required materials and minerals could increase by 500 percent by 2050. The scientists and authors of this article found renewable energy will add fuel to the fire when it comes to the threat that mining will do to our biodiversity.[184]

Just another small note in our big reveal. As our population increases, we'll need to feed it. There is a lesser-known gas that is released during the agricultural process—nitrous oxide, N_2O. The more we need to feed a growing population, the more N_2O will be released. And the gas is accumulating. No one really talks about it because we're fixated on CO_2 and cow farts. We need to have honest discussions on this subject. Maybe that's why gardeners are so happy.

My advice to the leftist media is that if you find yourself in a big hole, the first thing to do is stop digging!

I think we've said what needs to be said. What does all this information tell us? We have been manipulated, misinformed, and lied to. The current renewable resources that are being stuffed down our throats are about as efficient as putting a steering wheel on a donkey.

The list of conclusions tells us that all renewables need to be dumped or at least improved a great deal. We need to work on efficiency.

> **"I don't make jokes. I just watch the government and report the facts" (Will Rogers).**

8

Why Is This happening?

Yes, why is all this happening? We have all these renewables, and politicians are marching forward to the beat of the far left's drum. We will look at the money trail in this chapter as well as the Keystone Pipeline closing and how good electric vehicles are in terms of being powered by renewables. Most important is the massive increase in our population.

The Money Trail

Before we get into other matters, we're going to spend a little time talking about politics and money. Surely most of you have heard the phrase *follow the money*. And that goes for national and international politics and science as well. Following the money is like a lovesick goose following my cat. If you really want to see what is going on in any political or scientific arena, follow the money. Some say the only reason many renowned scientists are persuaded to agree that the apocalypse is at hand is that they are very well paid to say just that. Just putting out some opinions, folks.

"In America and around the globe, governments have created a multi-billion-dollar Climate Change Industrial Complex." Ever heard of it? Me thinks it's a real possibility some people are getting very, very rich.[171] Even richer than they already are. And they don't want that ball to stop rolling, and they don't want any of us to know about it.

Ideas have been thrown out there, so we, the people, can decide what we want done about all this hubble-bubble over climate change. Maybe it's all a government conspiracy theory, but money does strange things to people no matter ye' be liberal or conservative. Maybe it's a misinformation campaign to keep us down and in line. When you've got a lot of moolah, you want more and more of it, and you'll do about anything to get it. Money brings power.

Here is how climate research works. If research findings agree with the current apocalyptic view of climate change, then those researchers will receive billions of dollars from liberal governments and billionaires. If any researcher dares to question the results of the current doomsday climate change apocalypse scenario, they receive little research money and are accused of working for big oil and being racist. That's how big money corrupts science.

Here's the rub. Global spending has reached over $359 billion plus a year. The elites want trillions. A lot of that money, our tax money, goes to give subsidies to all these renewable high-tech industries. Why? Because they can't stand on their own.

Conclusion 1: with all this in mind, the US government and the UN admit, after decades and decades of research, "we have not delayed the apocalypse by a single day."[174]

That's because every single country on the planet has to be truly committed, and most are not. By the way, where has all that money gone? You'd think after all these years of pumping all that money into research and subsidies, they would have found some great tools to fight their climate change. There are no great tools as of yet.

Governments, including our own, are spending trillions of dollars to fight the weather, but they have failed. The public hasn't even asked the all-important questions because we all thought they were answered years ago, but they were not. What are these fundamental ideas the people never bothered to ask about but should have?

(1) Where is the *hard data* that *proves* that hydrocarbon energy, as in coal and oil, leads to temperature rise? Even with the very small temperature rise we do have, no one has produced any hard data that proves humans are the cause of it *all*. Like magic, it's a game of misdirection. That doesn't mean it isn't true; there's just no hard data that links the two. Or at least they haven't presented any.

(2) Who has put forth a "baseline climate ideal" for the planet? No one. There is no *ideal* temperature for the planet. The IPCC and other climate panels have laid out climate models galore. Yes, more than one. Many, in fact. They have told us we cannot go above a rise of 1.5–2.0°C (2.7–3.6°F) from preindustrial temperatures. Yet no one really knows what the preindustrial temperatures were. Not for sure, unless they are defining preindustrial as pre-1880. We only started recording accurate temperatures since the1880s. I've presented some *possibilities* about temperature rise. They have no idea what our cap might be, but they have to try convincing us that this giant con game is real because the elites have to continue making money.

(3) Who has produced a "cost-benefit analysis" of human life and standard of living needs to control the economy for the purpose of restructuring? No one. They hire ruthless individuals (I call magicians) to try to bring the people into the fold so they will give their money to *a good cause*. They won't mind higher taxes if it's to save the world. If this climate change industrial complex actually takes hold, this global warming con game will take power away from the people and hand it right over to those who would be king.[175]

But is it all true? Is it a huge, global magic show? A con game where a few are making millions upon millions of dollars? Or a conspiracy that there is no climate change? One thing for sure, the US isn't focusing on the real answers. Current renewables are not the answer.

They were never meant to be. Do the environmentalists go along with this or are they taken in as well?

From what I've read, "President Eisenhower, himself a former general, famously warned about the military-industrial complex and its potential hold on the U.S. in the late 1950's." With the huge changes in the world's political arena, our military complex has recently developed a new security threat ... climate change.[176]

We went from a military industrial complex in the 1950s to a new climate change industrial complex. So, are we headed for a military-climate-change industrial complex? The GND even touts that climate change is the new security threat. I can only assume the military industrial complex is working on strategies to cope with *climate change*. Answer this: what do you think a military complex will do if they perceive a global climate change threat is upon us? It's an interesting and terrifying thought. Think about it. I don't have the answer to that.

We've gotten that part done with, but what about the Green New Deal, politics, and money? Ever since the young socialist Alexandria Ocasio-Cortez and all her young followers stormed into Nancy Pelosi's office demanding something like the GND, the young in America rallied around her even though none of them, including Representative Cortez, knew anything about the issues. They were blindly following an ideal.

With left-wing young politicians just being elected to Congress, they had the enthusiasm and zeal that only the young could run with. They pushed their leftist, Marxist views onto their followers, using climate change as their calling card. The general premise in their politics was to reduce the economic value, at least in part, of the US resource base in energy-producing capital stock. Secretly. It would destroy our resources and increase the wealth of the already wealthy.[229]

Having put all this energy into their pet project of climate change, these politicians are in for a rude awakening. Even according to the IPCC reports, the result of the net-zero policies on future temperatures will be barely noticed. Reports of the IPCC and others agree that the effect would be 0.083°C by 2100, as compared to the 0.173 from zero by 2100. This is negligible as far as normal variations go. I can only conclude that the real agenda is wealth redistribution under the guise of a GND, as well as increasing government control. I think we are already seeing this.

This last point I want to make is that despite all this push for a 100 percent renewable energy system, we will still need a backup system in the case of renewable failure, and there's a lot of that going on. The backup comes in the form of natural gas. Here is the deal. I heard the governor of California had to import gas from the outside because he had a really bad failure of renewables. The president of the United States (Joe Biden) ended the Keystone Pipeline as well as oil leases on federal lands. After a while, those decisions didn't work so well, so he asked OPEC to increase their production. They said no. Shortly after that, a judge ruled that our president had no authority to end federal leases, and he is in the process of renewing leases. That's politics and money for you.

The Big Reveal

If I had the power to command the mitigation of CO_2e emissions, there would be a long list of must-dos. To start with, our first big reveal comes in the form of carbon. If you think about it, carbon is a miracle. What if we turn our thinking around for a minute? We view carbon as a villain. Carbon is the basis for all life on earth. It's the fourth most abundant element in the universe. You can do tons of stuff with it. Carbon atoms can be turned into diamonds. That would be cool. We already do that for industrial purposes. It can be formed into graphite, carbon fiber, and the weird nanomaterials. Why don't we use this carbon and reinvest it in steel and cement industries? Let's get scientists to start thinking about how all this can be done on a larger scale.

We are going through a heavy solar storm at this time. What's so important about that? If you remember, the IPCC scientists and others around the world have been telling us that the sun's cycles have no bearing on climate change or our weather, but they do affect our electrical grid. It is supposed to peak in 2025, so we are in it for a while.

Over the past decades, the IPCC has deliberately excluded and misrepresented important science regarding climate change. How do we know this? A group called the Nongovernmental International Panel on Climate Change (NIPCC) has challenged the IPCC on several fronts. Here is just one. I'm simply going to quote them so you can see the difference. The IPCC states, "Systemic risks due to extreme (weather) events leading to breakdown of infrastructure networks and critical services." On the other hand, the NIPCC reports,

> There is no support for the model-based projection that precipitation in a warming world becomes more variable and intense. In fact, some observational data suggest just the opposite, and provide support for the proposition that precipitation responds more to cyclical variation in solar activity.[215]

An astronomer and NOAA scientists have "now concluded that four factors determine global temperatures: carbon dioxide levels, volcanic eruptions, Pacific El Niño patterns, and the *Sun's activity.*" "*Global climate change*, including long-term periods of global cold, rainfall, drought, and other weather shifts, may also be influenced by solar cycle activity."[216]

So, what would I do if I had the power? Here's my list. Start with the US. Then try to expand this worldwide.

- Get rid of lobbyists. They are the ones preventing us from getting the real work done on temperature rise.
- Mobilize the planting of billions of trees worldwide. This is the first step because all the renewables destroy trees and the environment.

- Invest and demand better efficiencies, at least 50 percent or better, in our renewables. If the efficiencies were good, we wouldn't need so much land for wind and solar.
- Limit (by law) the amount of CO_2e that factories, mining, and processing plants can emit. Sounds easy. Lobbyists make it nearly impossible. They somehow get politicians to vote against anything that would limit big industry.
- Impose very high fines on those industries that fail to curb their emissions. Use those fines to invest in areas like carbon capture.
- Invest in carbon capture systems like Iceland's Carbfix projects, Canada's Modular Carbon Capture and Storage, and any other systems that sound promising. Make this a priority. The GNDers are screaming we are in an apocalypse. Let's get this done.
- Retrofit coal and natural gas plants as soon as possible with carbon capture systems. Not ten years from now. Yesterday would be preferable.
- Stop building more wind and solar until they provide much better efficiencies. Don't waste the land until these companies can produce more energy with high efficiencies.
- Use custom solutions where normal avenues are difficult to reach. Think outside the box.
- Repurpose some of that carbon.
- Make sure we are transparent and honest with the American people. Then we might not have such division in the country over climate change.
- I would ground John Kerry. He should be doing zoom lectures.

I'll leave you with two final thoughts. As renewables prove to be unreliable and ineffective, it appears that some environmentalists are turning their backs on our existing renewable energy sources. Their long-held beliefs that they were better for nature and ecosystems is melting away now that they realize the mining of materials is a greater threat to nature and our endangered species. A lot of the mining the world is engaged in is for renewable energy.

As the problems continue with renewables, German pronuclear environmentalists produced a banner in front of the Greenpeace headquarters that said, "Climate Crisis? Nuclear energy!" They protested all over Europe. Environmentalists finally realized that a uranium pellet, the size of a peanut, has as much energy as a ton of coal. What does that mean? You can make immense amounts of energy on a lot less land, thereby protecting a lot more of the environment.

We would not have to use more and more farmland for solar farms and wind farms. I'll give you a statement by Jonathan Ford, a columnist for *Financial Times*:

> It all goes to illustrate one of the awkward truths about renewables and one that is often buried beneath impressive statistics showing the declining costs of solar panels and wind turbines. Their relatively low 'power density' makes them more consumptive of resources.[227]

This final thought is a doozy. The report comes out of the UK but can represent any country, especially the United States. Politicians love passing laws to help in the reduction of greenhouse gases. In the UK, they plan to ban gas-fired domestic boilers to heat homes. Politicians are telling people they will transition to heat pumps. It's not surprising that they haven't put heat pumps in their homes.[228] They are super expensive and are pretty much trash.

People might take this seriously if it weren't for the fact that the secretary of state for business, energy, and industrial energy had to admit he doesn't have one in his own home. A politician has admitted he drives a diesel car and doesn't care for electric cars. The same goes for the US. We have climate gurus who think nothing of powering several huge mansions, flying all over the globe in private jets, lecturing on climate change.

Instead of putting trillions of dollars into projects we know will not work, put some of that money into research. Even if we don't find some new, great, innovative result, we will have learned a lot on the journey. With all this information that we have now, I think the United Nations said it best:

After decades of research and billions of dollars spent, "we have not delayed the apocalypse by a single day."

APPENDIX I

Wind Turbines

Formula 1: Efficiencies for a 2.5 MW rated turbine equals the number days/year X number hours/day X capacity X efficiency factor, using 365 days per year and 24 hours per day.

For a 2.5 MW capacity turbine at 100% efficiency (impossible):
365 days/year X 24 hours/day X 2.5 MW X 1(100%) = 21,900 MW hours/year

Calculation 1

Efficiency	days/yr X	hrs/day X	Capacity X	Efficiency Factor	= MWh/yr
100%	365	24	2.5	1.0	=21,900
75%	365	24	2.5	0.75	=16,425
50%	365	24	2.5	0.50	=10,950
35%*	365	24	2.5	0.35	= 7,665
25%	365	24	2.5	0.25	= 5,475

Data points on efficiency

*Average

Let's graph these efficiencies,

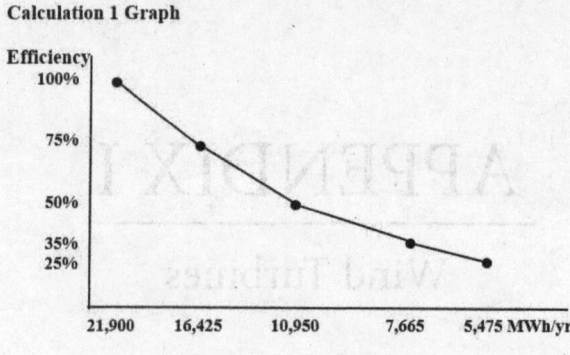

Graph from calculation 1

As long as the turbine works 365 days/year and 24 hours/day at maximum power, it will provide a lot of electricity. However, this is not an accurate measure of turbine output. From here on, the MWh/year takes a nosedive.[58]

Calculation 2:

No. days/yr	Hrs/day	Capacity	Efficiency Factor	MWh/yr
365	24	2.5	35%	= 7,775
292	24	2.5	35%	= 6,132
274	24	2.5	35%	= 5,754
245*	24	2.5	35%	= 5,145
182	24	2.5	35%	= 3,822

Data on efficiency by number of days per year

*Average

Notice the MWh/year goes down even more.

There is one calculation left in this series, and that is the average number of hours a turbine runs per day. Who the heck knows that? It so happens to be 16.4 hrs. You'll see why in a few minutes. Take formula 1 with the averages and the true efficiency factor, and this is what you get:

245d/y X 16.4h/d X 2.5 X 0.35(35% efficiency) = 3,515 MWh/year

This is the actual *average* MWh/year a 2.5 MW capacity turbine puts out per year.

Calculation 3: To convert MWh/year to MW/year:
simply divide the MWh/year by the number of hours a turbine runs per year.

3,515 MWh/year ÷ 6,000 hours/year = 0.59 MW/year

What! Is that right! A 2.5 MW capacity turbine only produces 0.59 MW of electricity a year. And where the hell did the 6,000 hours magically appear from? It comes from the European Wind Energy Association.[59] According to them, "Wind turbines can carry on generating electricity for 20–25 years." "Over their lifetime they will be running continuously for as much as 120,000 hours."

Here begin our calculations for determining how many MW/year a 2.5 MW capacity turbine produces per year. I know I already gave you the answer, but bear with me.

Calculation 4: Calculate the average number of hours per day:
120,000 hours ÷ 20 years (average) = 6,000 hours/year
6,000 hours/year ÷ 365 days/year = 16.4 hours/day

Calculation 5: Now we can convert MWh/year to MW:
MW = MWh/year ÷ number of hours a turbine runs per year.
Again, assuming optimal wind on all those days.
MW = Mwh ÷ hrs/year
1 year hours = 6,000

For our 2.5 MW tower: From formula 2, we determined the electrical output was 3,515 MWh/year. Our other parameter, is 6,000 hours/year following our formula:

3,515 MWh/year ÷ 6,000 hours applying the rules of math:

So, my friends, we come to the end of these calculations.

APPENDIX II

Calculations for Wind Turbines

Let's see what happens when turbines run fewer days, not 365. Use the same formula as in appendix I.

Parameters: runs 100% = 365 days/year, 24 h/d at 100% efficiency
365 x 24 x 2.5 x 1 = 21,900 MWh/year (the ideal but impossible)

But turbines don't ever work at 100 percent efficiency rating. They have to deal with friction and such. Research has determined, on average, a turbine has about a 33 percent efficiency rating. I'll use 35 percent like before.

Calculation 3: Change the number of hrs/day

No. days/yr	Hrs/day	Capacity	Efficiency Factor	MWh/yr
365	16.4	2.5	35%	= 5,238
292	16.4	2.5	35%	= 4,190
274	16.4	2.5	35%	= 3,932
245*	**16.4**	**2.5**	**35%**	**= 3516**
182	16.4	2.5	35%	= 2612
91	16.4	2.5	35%	= 1,306

*Average

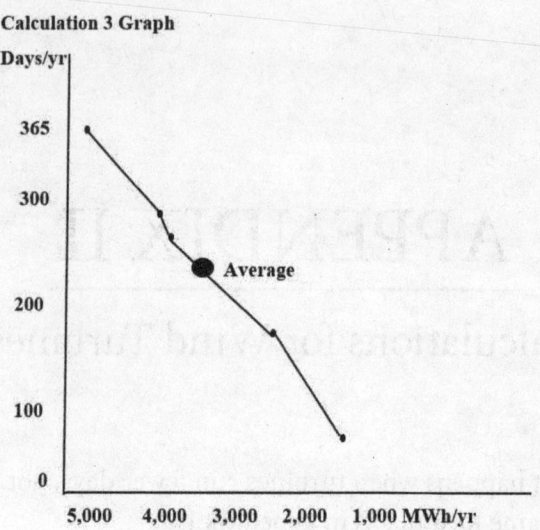

*Calculate 245 d/year X 16.4 hours/d X 2.5 X 0.35 = 3,515 MWh/year

This is an average MWh/year for what a 2.5 MW turbine might produce in a year. These are taking all the averages of how many days a year a turbine might run, how many hours per day a turbine might turn at an efficiency rating of 35 percent (which might be a tad high). This is all data I collected from my research on wind turbines. Again, you can see the numbers MWh/year keep dropping. There is a purpose to all this, I promise.

Let's see if we can make a graph out of this information.

Over a turbine's life, it will run on average 120,000 hours/20years[62]
= 6,000 hours/year = 6,000 hours/365 days = 16.4 hours/d = 16.4 hours/24 hours = 0.68th of a day.

0.68th of a day X 365 days/year = 248 d/year

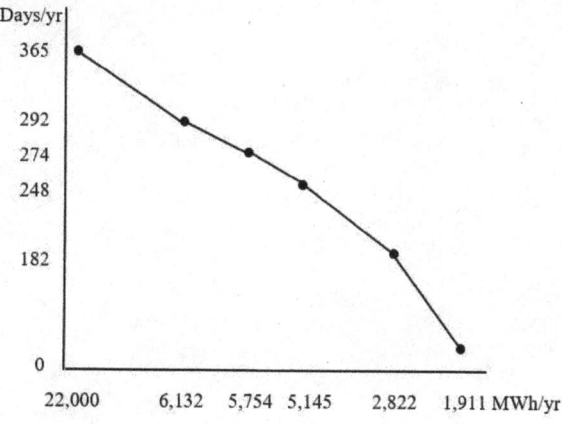

Days/year vs. MWh/year

APPENDIX III

MW a Turbine Produces per Year

Formula: MW = MWh/year ÷ Hours run/year

This might be visually easier:[61]

MW = MWhours ÷ Hours/year

Let's use the average numbers research has given us to see the average MW a 2.5 MW turbine is projected to produce.
Parameters: 245 days/year = 3,515 MWh/year (appendix II)
6,000 hours/year
0.35 efficiency rating

3,515 MWh ÷ 6,000 hours = 0.59 MW per year

Finally, we have arrived at an average of what a 2.5 MW capacity turbine might actually output in a year = 0.59 MW. *This is not an absolute but an average from the data I found.* Some turbines might be a little better, and as turbines age, they might run a little worse.

APPENDIX IV

Forests

Data: 1,900 grams of CO_2 are released for every 1,000 grams of timber.

Convert grams to pounds (lbs.).
1,900 grams = 4.2 lbs. 1,000 gr. = 2.2 lbs.
Therefore, 4.2 pounds of CO_2 are released for every 2.2 pounds of timber.

Continue with the math, and you get:
1.9 lbs. of CO_2 per 1 lb. of timber. Approximate ratio of 2:1

More simple math:
3,800 lbs. of CO_2 per 2,000 lbs. of timber

Keep working:
For every ton (2,000 lbs.) of timber, 1.9 tons of CO_2 are released. That's what we want.

Data: 8.8 million acres are burned in 2018 in the US.[13]
Average ton per acre = 87 tons/acre of timber

Calculation:
8.8 million acres X 87 tons timber/acre = 765.6 million tons of timber burned

From our earlier calculations: for every ton of timber burned, 1.9 tons of CO_2 are released.

765.6mt [million tons] timber X 1.9 tons CO_2/ton of timber = 1.5 billion tons of CO_2 was released in 2018 from forest fires… approximately.

APPENDIX V

Energy Production

The GND advocates removing all fossil fuels. The proponents dislike nuclear and other sources of energy. I can only assume that at some time in the future, these would be phased out. But certainly not by 2050. The only thing I have changed is getting rid of biomass burning altogether because it is *not* really a renewable.

Let's take a look at 2019.

2019 production: 101 quads total energy production = 3,379,016 MW/year

    Nuclear – 8.5 quads = 284,373 MW/year
    Natural Gas – 34.9 quads = 1,167,601 MW/year
    Petroleum – 31.8 quads = 1,063,888 MW/year
    Coal – 14.3 quads = 478,415 MW/year

Renewables are 11.6 quads of the total energy production = 388,085 MW/year

    2% - Geothermal   = 7,762 MW/year
    22% - Hydroelectric   = 85,379 MW/year
    24% - Wind   = 93,140 MW/year
    9% - Solar   = 34,928 MW/year
    43% - Biomass   = 166,877 MW/year[13]

If we keep to the ideals of the GND, we will have 100 percent of our energy renewable. We'll put a little scenario together where, by 2050, we will be 92 percent renewable. The Energy Information Administration predicts we will still use nuclear, so let's keep that in the mix. Yeah for the GND! How do we break this down into percentages?

Energy projection for production 2050: 117 quads = 3,914,305 MW/year

Using this scenario, my projection for 2050 would be:

7.8% nuclear
0.2% geothermal (stays the same)
Equation: $2\%/x = 11\%/100\%$; $02/0.11 = 0.2\%$
2% hydroelectric (static mature state)
Equation: $22\%/x = 11\%/100\%$; $0.22/0.11 = 2\%$
37% wind
53% solar
0% biomass (I hate biomass)

We'll start here and see where this takes us. I removed biomass because it is a very dirty industry, and I don't think it belongs in our renewable list. I may have to backtrack on that a little and have to keep it but only at its 2019 level. So I'll stick it in even though I'd prefer to get rid of it like a rotten egg.

2050 Energy Projection

Projection = 117 quads = 3,914, 305 MW/year

7.8% Nuclear	= 282,900 MW/yr (same as 2019)
0.2% Geothermal	= 7,762 MW/yr (same as 2019)
2% Hydroelectric	= 85,379 MW/yr (same as 2019)
4.5% Biomass	= 171,962 MW/yr (same as 2019)
TOTAL	376,041 mw/yr
Renewables 35% Wind 51% Solar	= 1,370,007 MW/yr = 1,996,295 MW/yr
TOTAL	= 3,538,264 MW/yr

We could do all kinds of different percentages, but we're not going to. Where are we going to be in 2050? Environmentalists don't much like nuclear or hydroelectric and aren't too fond of geothermal. One article projects biomass production for 2050 will be around 5.39 quads or 180,326 MW/year.[219] It's near impossible to estimate what the future will be.

I found an article by the US Energy Information Administration that projects power generation for 2050. It's a pretty realistic mix. Renewables will be at 42 percent, with wind and solar responsible for most of it.[220] It expects nuclear and coal to decrease a bit and natural gas to stay about the same. This sounds more reasonable. Unless, of course, they find out there is no *scary* climate change. In any case, we need to control our CO_2e emissions.

If, for 2050, the wind power production will be about 1,370,007 MW/year, then we need to figure out how many wind turbines we actually need. In chapter 2, we calculated that a 2.5 MW tower only produces 0.6 MW/yr. How many turbines do we need to produce 1,370,007 MW/year?

(1 Turbine/0.6mw/year = x turbines/1,370,007mw/year) = 1,370,007÷0.6

It turns out to be 2,283,345 turbines.

How much land do we need? 0.25mi^2 per turbine, so

2,283,345 turbines x 0.25 mi²/turbine = 570,836 mi² approximately

We can do the same thing for solar. It also comes out to be a lot of square miles.

Just a little reminder: California has 163,696 mi².

You're going to tell me we won't need that many turbines because we are also using solar. I get that so let's cut the land needed by half:
570,836 ÷ 2 = 285,418 mi²

It's still bigger than the state of California.

APPENDIX VI

Population, Number of households, CO_2, Methane, SF_6

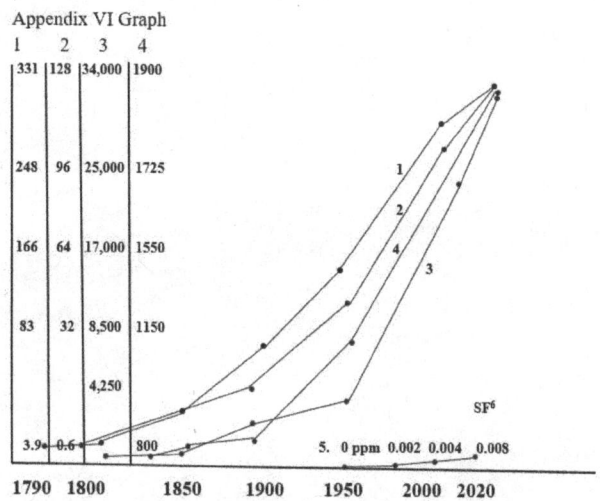

1. US population—millions
2. Number of US households—millions
3. CO_2 emissions—million metric tons
4. Methane emissions—PPB
5. SF_6 emission—PPB[177]

FIGURES and GRAPHS

Figure 1: Graph of US hurricanes 26
Figure 2: Graph of tornadoes 28
Figure 3: Tornadoes by year 29
Figure 4: Annual F3+ tornadoes 30
Figure 5: Blizzards versus time 31
Figure 6: Data points for figure 7 48
Figure 7: Acres burned over time 49
Figure 8: Nuclear versus wind power 60
Figure 9: Steel production 92

Endnotes

1. Robinson Meyer, "The Think Tank Struggling to Write the Green New Deal," *Atlantic* (June 12, 2019).
2. NASA, "Overview: Weather, Global Warming and Climate Change," (n.d.).
3. United Nations, Department of Economic and Social Affairs, "Growing at a slower pace, world population is expected to reach 9.7 billion in 2050 and could peak at nearly 11 billion around 2100," (June17, 2019), https://www.un.org/development/desa/en/.
4. "How Do Trees and Forests Relate to Climate Change?" *Guardian* (February 11, 2011).
5. NASA, "Global Climate Change. Vital Signs of the Planet. The Effects of Climate Change," (n.d.).
6. UCAR center for Science Education. "Carbon Dioxide Absorbs and Re-emits Infrared Radiation," (2012).
7. EDF, Environmental Defense Fund. "How Climate Change Plunders the Planet," (2020).
8. NASA, "Observations from Space, Sea Level Change. How long have sea levels been rising? How does recent sea levels compare to that over the previous centuries?" (n.d.).
9. National Park Service, "Wildfire Causes and Evaluations," Wildland Fire Management Information and U.S. Forest Service Research Data Archive, (2000-2017). (n.d.).
10. Libby Emmons, "Australia arrests 183 for setting brushfires that celebrities claim were caused by climate change," *The Post Millennial, Canadian News*, (2019).
11. Verisk's 2019 Wildfire Risk Analysis. "Facts + Statistics: Wildfires," (2019).
12. Nathan Thomas, "How many tons of wood are on an acre of land?" Forest2Market, (August 13, 2018).

13 U.S. Energy Information Administration (EIA), "U.S. Energy facts explained – consumption and production."
14 Mark Anderson, "Something Not Right with CA Wildfires," *American Free Press*, (December 8, 2019)
15 James Boettcher, "How many people still use wood fire for heating/cooling *worldwide*?" IT Consultant, (March 31, 2016).
16 Hugh Willoughby, "Better Hurricane observation techniques over decades make big storms less deadly", PHYS.ORG, Earth Sciences, The Conversation, (June 2, 2015).
17 Dennis Mersereau, "At 200 mph, Hurricane Patricia is now the strongest tropical cyclone ever recorded, *The Vane*, (October 23, 2015).
18 Wikipedia, "Great Hurricane of 1780".
19 Stormfax Weather Almanac, "Atlantic Hurricane Numbers by Year: 1851-2017," Original Data from The Deadliest, Costliest, and Most Intense United States Hurricanes 1900-2000.
20 NOAA, "Historical Records and Trends," Climate Information, US Tornado Climatology.
21 Wikipedia, *Fujita Scale*.
22 Weather Almanac, vol. 1, "Tornadoes. Annual official total of tornadoes by year: 1953-2001".
23 History Stories, "Major Blizzards in US History," (November 29, 2018).
24 Charles Q Choi, "How Weather Satellites Changed the World," Earth Networks, (April 13, 2010).
25 Wikipedia-Blizzard, "List of Blizzards".
26 Chelsea, Harvey, "Love Snow? Here's How It's Changing," *Scientific American*, (January 28, 2019).
27 Doyle Rice, "Researcher: Number of blizzards doubled in past 20 years," (January 24, 2016).
28 The Earth Institute Columbia University, "2,000 Years of Climate in Europe," New Drought Atlas. (November 6, 2015).
29 Douglas Broom, "5 Droughts That Changed Human History," World Economic Forum Annual Meeting, (May 27, 2019).
30 Bjorn Carey, "Sahara Desert Was Once Lush and Populated," Live Science, (July 20, 2006).
31 Jesse Greenspan, "7 Withering Droughts," History, (August 22, 2018).
32 Wikipedia, "Droughts in the United States".
33 Worldometer, "World Population by Year," www.worldometer.info
34 B. Schildgen, "How much CO_2 is generated by producing and transporting a gallon of gas?" National magazine of the Sierra Club. Hey Mr. Green! (January 8, 2017).

35 G. Maryland and R. J. Andres, "Man-made (anthropogenic) greenhouse gases," Ref. T. A. BGS. British Geological Society. Global, Regional, and National Fossil-Fuel CO_2 Emissions. Carbon Dioxide Information Analysis Center, Oak Ridge National Laboratory, Department of Energy, (2009).
36 Wikipedia, "Chlorofluorocarbon".
37 Wikipedia, "Fluoroform".
38 MIT Technology Review, "There's been a huge spike in the world's most potent greenhouse gases," www.technologyreview.com
39 A. King, B. Henley, E. Hawkins, "What is Preindustrial climate and why does it matter?" *The Conversation*, (June 7, 2017).
40 B. Tisdale, "What was Earth's Preindustrial Global Mean Surface Temperature, In Absolute Terms Not Anomalies, Supposed to Be?" What's Up with That, (November 27, 2018).
41 E. J. Ritchie, "Exactly How Much Has the Earth Warmed? And Does it Matter?" lecturer, Dept. of Construction Management, (September 7, 2018).
42 Arctic News, "Temperature Rise From 1750 to 2016," Climate Plan-concl.
43 Liz Osborn, "History of Changes in the Earth's Temperature," Current Results, weather and science facts.
44 NOAA, "Global Climate Report," National Centers for Environmental Information. (Annual 2017).
45 Renee Cho, "Could Climate Change Shut Down the Gulf Stream?" Earth Institute Columbia University, State of the Planet, (June 6, 2017).
46 Eric Hand, "Crippled Atlantic Currents Triggered Ice Age Climate Change," *Science Magazine*, (June 30, 2016).
47 Erica Langston, "In 20 Years, Wildfires Will Be Six Times Larger," (May 22, 2018). www.Outsideonline.com
48 Congressional Research Service, "Wildfire Statistics," (October 3, 2019).
49 National Interagency Fire Center (NIFC), "Total Wildland Fires and Acres (1926-2018)".
50 National Centers for Environmental Information (NCIE), "Wildfires – Annual 2019," (March 6, 2020)
51 Fred Stutzenberger, "Fire-Shaping the New World Forests by Native Americans," Muzzleloader magazine, (vol. November/December 2006).
52 John Weier, "Mapping the Decline of Coral Reefs," NASA, World Observatory, (March 12, 2001).
53 SECORE, "Why Coral Reefs Need Our Help".
54 Wikipedia, "Cruise Ship Pollution in the United States".
55 Wikipedia, "Cold and heat adaptations in humans".

56   Lisa Friedman, "Heat Stress Drives Climate Migration," Climate Wire. *E&E News*, (January 27, 2014).
57   Herzog, Howard, and Smekens, "Special Report Carbon Capture and Storage, Cost and Economic Potential," SRCC.
58   Kevin Lee, updated by, "How Much Power Does a Wind Turbine Generate?" (April 24, 2018). https://sciencing.com
59   National Wind Watch, "Output from industrial wind power".
60   Univ. of Michigan, "Wind Energy Factsheet," css.umich.edu/factsheets/wind-energy-Factsheet
61   Morgan Owens, "How to Convert Megawatt Hours to Megawatts," (July 17, 2017). https://www.hunker.com
62   EWEA (European Wind Energy Association), "Wind Energy's frequently asked questions (FAQ)".
63   Carmel Pule', "How much energy does a 2.5-3 MW wind turbine produce an hour?" (July 20, 2019). www.quora.com
64   Rosenbloom, Eric Rosenbloom, "Size Specifications of Popular Wind Turbine Models," (November 1, 2005). www.windAction.org
65   Fullfact, "Overblown: Wind Turbines don't take more energy to build than they will ever produce," (February 27, 2019). https://fullfact.org
66   Stop These Things, "The Truth About the Great Wind Power Fraud," (August 16, 2014). https://stopthesethings.com
67   Paul Dvorak, "Take a Closer Look at Pouring Turbine Foundations," (December 31, 2012). www.windpowerengineering.com
68   Isaac Orr, "So You Want Wind Turbines But You Don't Want Copper Mines," (April 11, 2018). www.americanexperiment.org
69   Copper Development Association, Inc. Copper Alliance, "Copper's Role in Wind Generation".
70   "Size of Industrial Wind Turbines," www.wind-watch.org
71   Institute for Energy Research. IER. "Big Wind's Dirty Little Secret: Toxic Lakes and Radioactive Waste," (October 23, 2013).
72   "How Steel is Made: a Brief Summary of a Blast Furnace". www.keenovens.com
73   "How Much Coke Does it Produce from 1 Ton of Coal?" www.quora.com
74   "Steel: Raw Materials". www.science.jrank.org
75   Kundak, L. Lazic, and J. Crnko, "CO_2 Emissions in the Steel Industry," *Metalurgua*, (2009).
76   "How Much CO_2 Gets Emitted to Build a Wind Turbine?" (August 16, 2014). https://stopthesethings.com
77   "Frequent Questions About Coal Mine Methane," (November 14, 2018). www.epa.gov

78. Portland Cement Association, America's Cement Manufacturer, "How Cement is Made," (2019).
79. The Green Ration Book, "Carbon Footprint of Concrete," (2010).
80. Isabel Van Driessche, "Life Cycle Assessment of Completely Recyclable Concrete," (February 14, 2016). www.researchgate.net
81. Resourceful Paths, "Sustainability in Mining-Blog," (November 23, 2016).
82. The Nautilus Minerals Inc, "The Long-Term Liability of Copper mining," Solwara 1 Project.
83. TENORM, "Mining and Production Waste," (July 8, 2019). www.epa.gov
84. FAQ-SIZE, nationalwindwatch.org
85. Oliver Vidal, "Mineral Resources and Energy," Nacelle. (2018). https://www.sciencedirect.com/topics/earth-and-planetary-sciences/nacelle
86. *Associated Press*, "Largest Wind Turbine Firm Dedicates U.S. Plant," (March 9, 2008). www.nbcnews.com
87. Michelle Froese, "How are blade materials and manufacturing changing to keep up with larger turbines," (January 4, 2017). www.windpowerengineering.com
88. Greg Petsche, "Blade Runners: A Look Inside a Factory for Giant Wind Turbine Blades," (June 11, 2018). www.ge.com/reports
89. "Emissions Factors in kgCO2 per unit," Winnipeg.ca.
90. G. Tchana, Toffe, et al., "A Scale-up of Energy-Cycle Analysis on Processing Non-Woven Flax/PLA Tape and Triaxial Glass Fibre Fabric for Composites," Journal of Manufacturing and Materials Processing, (November 1, 2019).
91. Jared Paben, "Company expands wind turbine recycling operation," Plastics update, (March 27, 2019). https://www.resource-recycling.com/plastics
92. Christine Peterson, "Rare earth processing: a complicated proposition," (September 27, 2014). https://trib.com/business,energy
93. "Rare-earth element – Processing Ores". https://www.britannica.com,science
94. Vogel, Hanno et al., "Reducing Greenhouse Gas Emission from the Neodymium Oxide Electrolysis. Part I: Analysis of the Anodic Gas Formation," Crossmark, (October 14, 2016).
95. Steve Blankinship, "Keeping Wind Turbines Spinning," Power Engineering, (issue 8, vol. 112), (August 1, 2008).
96. Justin Martino, "Renewable Energy World. Wind Turbine Lubrication and Maintenance: Protecting Investments in Renewable Energy," (May 21, 2013).

97. "How Much Does a Barrel of Oil Weigh?" https://wwwreference.com/science
98. D. Gordon, A. Brandt, J. Bergerson, and J. Koomey, "Know Your Oil, Creating a Global oil-Climate Index," Carnegie Endowment for International Peace, (2015).
99. *epa*, "Greenhouse gas reporting program," (2018). www.epa.gov/ghgreporting
100. T. Wang, Publisher, "U.S. refinery facilities' GHG emissions 2011-2018," (November 7, 2019). https://www.statista.com
101. Eric Beckman, "The World of Plastics, in Numbers," University of Pittsburgh, (August 9, 2018). https://theconversation.com
102. Steve Pryor, "A Partial list of over 6,000 products made from one barrel of oil (after creating 19 gallons of gasoline)". www.linkedin.com
103. Blog: Innovative Wealth, "144 Products Made From Petroleum and 4 That May Shock You". https://www.innovativewealth.com
104. Matt McGrath, "Climate Change: Electrical Industry's 'Dirty Secret' Boosts Warming," (September 13, 2019). www.bbcnews.com
105. Dragan Bogunovic, "Equipment CO2 Emission in Surface Coal Mining," Engineering, International Journal of Mining and Mineral, (January 2009). https://researchgate.net
106. Quora.com, "Which metal is Used in Solar Cells?" (2016 & 2018).
107. "What are Solar Cells Made Up of?" (2020). www.renewableenergyhub.co.uk
108. Minerals Database. https://mineralseductioncoalition.org
109. Greenfacts, "Boron: 2. Where is boron found". https://www.greenfacts.org
110. A. Stamp, P. A. Wäger, & S. Hellweg, "Science for Environment Policy," (Indium), European Commission. Issue 435, (November 12, 2015).
111. Periodic Table of the Elements, (Gallium). https://mineralseductioncoalition.org
112. Wikipedia, "Selenide". (May 26, 2019)
113. General Kinetics, "Tin Mining & Processing: Everything You Need to Know".
114. Wikipedia, "Zinc Mining," (2020).
115. General Kinetics, "Sulfur Mining & Processing: What to Know".
116. "Using Sulfur to Store Solar Energy," (May 2017). https://phys.org
117. Wikipedia, "Silver Mining".
118. Wikipedia, "Silicone Rubber".
119. Wikipedia, "Ethylene".
120. By the editors of Encyclopedia Britannica, "Polyvinyl Acetate". (March 9, 20020). https://britannica.com

121 Wikipedia, "Polyethylene terephthalate".
122 "The Solar Power Paradox: Alternative Energy Can't Run on Oil". (2009) https://www.altenenergymag.com
123 Venkatesh, "How is Aluminum Made?" https://www.scienceabc.com
124 "Solar Panel Radiation – The Complete Guide". https://emfacademy.com
125 Solar Energy Industries Association, retrieved from original article, "Concentrating Solar Power," (2017). https://www.seia.org
126 Wikipedia, "Ivanpah Solar Power Facility".
127 Jason Deign, "America's Concentrated Solar Power Companies Have All but Disappeared," (January 20, 2020). https://www.greentechmedia.com
128 David Laine, "Effects of Solar Power Farms on the Environment," (April 24, 2017). https://sciencing.com
129 Abi Grogan, "Steel and Steam Form Reliable Partnership," (September 2017). https://www.worldsteel.org
130 Argonne National Laboratory, "Life-Cycle Analysis Results of Geothermal Systems in Comparison to Other Power Systems".
131 "Geothermal Energy Pros and Cons," blog, (2015). https://www.comfort-pro.com
132 Bureau of Reclamation/Hoover Dam, "Frequently Asked Questions".
133 EIA, "Hydropower and the Environment," (2020). https://www.eia.gov/energyexplained/hydropower
134 John Walden, "Blocked Migration: Fish Ladders on U.S. Dams Are Not Effective," (April 13, 2013). https://e360.yale.edu/features
135 EnergySage, "Environmental Impacts of Hydropower," (September 27, 2019). https://www.energysage.com
136 Steven Bushong, "How are Ocean Waves Converted to Electricity?" (April 19, 2017). https://www.energy.gov/eere/articles
137 "How Does Wave Energy Work?" (February 5, 2019). https://www.surfertoday.com/environment
138 "Environmental Impact of Wave Energy Devices at Sea". https://www.alternative-energy-tutorials.com
139 William Hubbard, Chapter 4-"Wood Bioenergy," (2015). https://www.sciencedirect.com/science/article
140 J. Sterman, Lori Seigel, and Juliette Rooney-Varga, "Does Replacing Coal With Wood Lower CO_2 Emissions? Dynamic Lifecycle Analysis of Wood Bioenergy," (January 18, 2018). https://iopscience.iop.org
141 Michael Moore Presents "Planet of the Humans", video, directed by Jeff Gibbs.
142 Willem Post, "Do Wood-burning Power Plants Make Sense?" (April 16, 2015). https://www.greenbuildingadvisor.com/article

143 Ines Hajdu, "The Best Practices for Using Plant Residues," (2016). https://blog.agrivi.com/post

144 GreenFacts, Facts on Health and the Environment, "Forests & Energy. How is bioenergy produced?" (2008). https://www.greenfacts.org/en/forests-energy

145 J.D. Van Zyl. "Steel Helps Power the Vehicles of the Future," (June 2019). https://stories.worldsteel.org/automotive

146 SANDAG, Center for Sustainable Energy, "Electric Charging Station Installation Best Practices," (June 2016).

147 Wikipedia, "Cobalt".

148 Nellie Peyton, "Electric Car Demand Fueling Rise in Child Labor in DR Congo: Campaigners," *Reuters*, (November 2, 2018). https://reuters.com/article

149 Stephan Sabo-Walsh, "The Hidden Risks of Batteries: Child Labor, Modern Slavery, and Weakened Land and Water Rights," (March 19, 2017). https://wwwgreentechmedia.com/articles

150 Rebecca Bertram, "Is Latin America's Lithium Industry Sustainable? Environmental Costs of the New White Gold," (June 5, 2019). https://energytransition.org

151 Amit Katwala, "The Spiraling Environmental Cost of Our Lithium Battery addiction," (August 5, 2018). https://www.wired.co.uk/article

152 Roger Pielke, "Net-Zero Carbon Dioxide Emissions By 2050 Requires a new Nuclear Power Plant Every Day," (September 30, 2019). https://www.forbes.com/sites/rogerpielke

153 John Correia, Ansnuclearcafé.org. American Nuclear Society. "Wind Power to Nuclear Power Infographic Comparison," (February 9, 2012).

154 Wikipedia, *"Economic Transformation"*.

155 Study Introduction, Simply Insurance, "Average Life Expectancy in the U.S. by State, Gender & Age," (2020)

156 Nicholas Kusnetz "U.S. Emissions Dropped in 2019: Here's Why in 6 Charts," (January 7, 2020). https://insideclimatenews.org

157 The White House, "Findings from Select Federal Reports. The National Security Implications of a Changing Climate," (May 2015).

158 Blog. Glosbe.com, "Threat Multiplier," definition – English.

159 UN General Assembly, Report to the Secretary-General. "Climate Change and its Possible Security Implications," (September 11, 2009).

160 Dictionary.com., "Adaptation,"

161 Open access governments, "How Does Economic Development Impact Climate Change," (July 31, 2020).

162 Graeme Stuart, "What is Community Building," Word Press, (March 10, 2014). https://sustainingcommunity.wordpress.com/2014/03/10/ccb
163 S. Calle, and D. Bailey, "Earth Science Communications," NASA. Science Edit, (Updated August 4, 2020).
164 Adam Day and Jessica Caus, "Conflict Prevention in the Era of Climate Change: Adapting the UN Climate-Security Risks," (March 31, 2020). https://cpr.unu.edu/climate security.html
165 Wikipedia, "Community Resilience," (updated September 6, 2019). https://en.wikipedia.org/wiki/Community_resilience
166 Dept. of Energy, "Heat Pump Systems".
167 Ad. "How Heat Pump is Made".
168 The Environment, "Why are Hydrocarbon Bad for the Environment? https://the-environment.org.uk/further_info/hydrocarbons
169 Minnesota Dept. of Agriculture, "Ecological Effects of Ammonia". https://www.mda.state.mn.us/ecological-effects-ammonia
170 J. Mulligan, G. Ellison, K. Levin, "6 Ways to Remove Carbon Pollution from the Sky," World Resource Institute, (September 10, 2018). https://www.wri.org/blog.2018/09/6-ways-remove-carbon-pollution-sky
171 Robin Young and K. Miller-Medzon, "Researchers in Iceland Can Turn CO2 into Rock," *Could it Solve the Climate Crisis?"* Wbur Here & Now, (December 10, 2019). https://www.wbur.org/hereandnow/2019/12/10/iceland-climate-change-carbon
172 Madison Dapcevich, "Something Living at the Bottom of the Sea is Absorbing Large Amounts of the CO2 in Oceans," Ifl science, (November 21, 2018). https://www.iflscience.com/environment
173 Michael Irving, "Carbon Reservoirs in the Ocean Floor May have Ended the Last Ice-Age and Could Bubble Up Again," New Atlas, (February 15, 2019). https://newatlas.com/seafloor-carbon-dioxide-reservoirs-ice-age/58473/
174 Stephen Moore, "Follow the (Climate Change) Money.," The Heritage Foundation. Commentary, (December 18, 2018). https://www.heritage.org./environment/commentary/follow-the-climate-change-money
175 J.B. Shurk, "What the global warming advocates really have in mind," American Thinker, (Jan. 13, 2020).
176 David Biello, "Beware the military-industrial complex and climate change," News Blog, (March 28, 2008). https://blogs.scientificamerican.com/news-blog
177 Census Charts, "Chart of US Population, 1790-2000". https://census-charts.com

178 Ethan Siegel, Ask Ethan: "Will Earth's Temperature Start Decreasing Over the next 20,000 Years?" (October 16, 2020). https://forbes.com/sites/startswithabang/2020
179 Harvey Rice, "Rise in Sea Level Makes Hurricanes Worse." *Houston Chronicle*, (May 30, 2017). https://www.govtech.com
180 Ellen R. Wald, "The U.N. Says America is Already Cutting So Much Carbon It Doesn't Need The Paris Climate Accord," Former contributor *Forbes*, (December 10, 2020). https://www.forbes.com
181 J. Benson, "Green Debacle – Tens of Thousands of Abandoned Wind Turbines Now Litter American Landscape," (November 24, 2011). https://www.bibliotecapleyades.net
182 David Knowles, "Extreme winter storms aren't inconsistent with global warming and will continue for decades, expert says," Editor *Yahoo News*. https://news.yahoo.com
183 Kurz industrial solutions, "Spacing it out: How much land is required for wind turbines?" (January 21, 2019). https://kurz.com
184 Laura Sonter, J. Watson and R. Valenta, "A Vast Transition From Fossil Fuels to Renewable Energy is Crucial to Slowing Climate Change," Inverse, The Conversation, (September 6, 2020). https://www.inverse.com/science
185 Des Moines, Iowa (AP). "MidAmerican idles 46 wind turbines after blades fall from 2 towers," (October 21, 2020).
186 Paul Dvorak, "Vertical-axis wind turbines: what makes them better?" Wind Power Engineering, (October 31, 2014). https://windpowerengineering.com
187 Solar Reviews Blog. "How long do solar panels actually last?" (October 22, 2020).
188 Liam Stoker, "Built solar assets are 'chronically underperforming' and modules degrading faster than expected, research finds", PV-tech news, (June 8, 2021). https://www.pv-tech.org
189 Michael Barnard, "The Truth About Efficiency in Solar Power Generation," Contributor on Quora, *Huffington Post*, (November 8, 2017). https://www.huffpost.com
190 Joe Clements, "Solar Farm Land Requirements: How Much Land Do You Need?" Greencoast, (June 19, 2019). https://greencoast.org
191 Wikipedia, "Solar Star".
192 Questions on Solar Power, "The biggest solar farm in the United States".
193 Garrett Hering, "4 reasons the Ivanpah plant is not the future of solar," (February 19, 2014). https://www.greenbiz.com/blog
194 Wikipedia, "Concentrated solar power".
195 National Park Services – Mojave "Desert Tortoise," (December 21, 2020). https://www.nps.gov

196. Jason Deign, "America's Concentrated Solar Power Companies Have All but Disappeared," Greentech Media, (January 20, 2020). https://www.greentechmedia.com
197. Pam Boschee, "On Becoming Obsolete: How High-Tech Solar Plant Found Its Way to Bankruptcy," JPT, (August 3, 2020). https://jpt.spe.org
198. A. Calderon, C. Barreneche, et al., "Review on solid particle materials for heat transfer fluid and thermal energy storage in solar thermal power plants," Abstract. online library. (May 15, 2019). https://onlinelibrary.wiley.com
199. Clifford Ho, "Concentrating Solar Power and Thermal Energy Storage," Sandia National Laboratories.
200. Partnership for Policy Integrity, "Carbon Emissions from Burning Biomass for Energy," info@pfpi.net
201. Environmental Law Alliance Worldwide, "Health impacts of open burning of used (scrap) tires and potential solutions". https://www.elaw.org
202. Mark Mills, "If You Want 'Renewable Energy,' Get Ready to Dig," Global Warming Policy Forum, (June 8, 2019). https://www.thegwpf.com
203. Joanna Marsh, "What Keystone pipeline cancellation means for crude-by-rail," Freightwaves. (February 15, 2021). https://www.freightwaves.com
204. Nia Williams, and D K. Kumar, "Even without Keystone XL, set for record Canadian oil imports," (January 22, 2021). https://www.reuters.com
205. "Oil and Natural Gas Pipelines," 2020.
206. Michael Rubin, "Biden kills pipelines at home but promotes them for the Taliban," *Washington Examiner*, (February 8, 2021).
207. Chris Mooney, "An enormous missing contribution to global warming may have been right under our feet," *Washington Post*, (June 4, 2021).
208. Published by N. Sönnichsen. "Energy consumption and production in the United States from 2020 to 2050," (February 22, 2021). https://www.statista.com
209. Simon Lewis, "Sucking carbon out of the air is no magic fix for the climate emergency," *Guardian*, (August 1, 2019). https://www.theguardian.com
210. Simpson Strong-Tie. Blog. "Hydrogen Embrittlement in High-Strength Steels".
211. Richard Black, "Eliminating Emissions," Director, Energy and Climate Intelligence Unit. (April 27, 2020).
212. Catherine Clifford, "Carbon Capture Technology Has Been Around for Decades – Here's Why It Hasn't Taken Off," (February 1, 2021). https://www.cnbc.com
213. R. Sigurdardottir, A. Rathi, and Bloomberg, "This startup has unlocked a novel way to capture carbon – by turning the foul gas into rocks," (March 6, 2021). https://fortune.com

214 Advantage Oil & Gas Ltd, "Advantage Announces Advanced Modular Carbon Capture and Storage ("MCCS") Technology, First Commercial MCCS Deployment at Glacier, and Founding of Entropy Inc." (March 31, 2021).

215 Avik Roy, "The IPCC's Latest Report Deliberately Excludes and Misrepresents Important Climate Science," Opinion Editor, Capital Flows, Forbes opinion, (March 31, 2014). https://www.forbes.com

216 Catherine Boeckmann, "What Are Solar Cycles, And How Do They Affect Weather," Old Farmer's Almanac, (March 4, 2021).

217 Michael Shellenberger, "The Reason Renewables Can't Power Modern Civilization Is Because They Were Never Meant To," Contributor, *Forbes*, (May 6, 2019). https://www.forbes.com

218 Michael Bastasch, "IPCC Report: Extreme Weather Events Have Not Increased," *Daily Caller*, (August 10, 2018). https://www.thegwpf.com

219 U.S. Energy Information Administration (EIA). "EIA projects renewables share of U.S. electricity generation mix will double by 2050," (February 8, 2021). https://wwweia.gov/todayinenergy

220 Madhumitha Jaganmohan, "Biomass energy production forecast in the U.S. 2020-2050," (February 22, 2021). https://www.statista.com

221 "White House Initiative on Global Climate Change, Climate Change over the Past 100 Years. (1990's).

222 UCR, NASA, RUSD, "Down to Earth Climate Change, Teaching tomorrow's leaders today".

223 *IBERDROLA*, "How is climate change affecting the economy and society?" Article.

224 Joanna Thompson, "Atlantic Ocean Currents Weakening, Near Verge of Collapse, Study Says," How stuff works, (August 24, 2021).

225 Giles Parkinson, "Bomen solar farm collects damages payment after three week outage," (August 24, 2021). https://reneweconomy.com.au

226 Population Matters, "Climate Change".

227 Michael Sellenberger, "As Renewables Falter, Environmentalists Stand Up For Nuclear," (September 9, 2020).

228 Rob Lyons, Columnist, "Even politicians don't want to give up their gas boilers," (August 25, 2021).

229 AEI, American Enterprise Institute, "The Green New Deal: Economics and Policy Analysis," (March 26, 2019).

230 Andrew Lisa, "History of droughts in the US," Stacker, (March 5, 2021).

Index

100 percent of our energy needs 75
access to clean water 74
access to healthy food 78
achieve net-zero 72
achieve the GND will require 83
big reveal 153, 166, 173
biomass 2, 24, 37, 81, 117, 121, 122, 123, 126, 129, 136, 137, 138, 139, 154, 155, 163, 164, 191, 192, 193
Biomass 136
BIOMASS 136
Boron 108, 204
Cadmium telluride 108
Carbon Capture 157
CHANGING CLIMATE 17
Cleaning existing hazardous waste sites 82
cleaning up pollution 78
climate change v, ix, xi, 1, 2, 3, 4, 5, 6, 7, 8, 9, 11, 12, 13, 14, 15, 16, 18, 19, 20, 21, 22, 24, 25, 32, 41, 42, 44, 46, 49, 51, 53, 56, 57, 60, 61, 62, 63, 64, 65, 66, 67, 68, 69, 70, 71, 73, 75, 79, 82, 83, 125, 127, 136, 145, 147, 148, 150, 152, 153, 157, 162, 165, 170, 171, 172, 174, 175, 176, 193, 199, 207, 210
Climate Change Industrial Complex 169
Community Resilience 73
community resiliency 73
concentrated solar power 111, 112, 133, 163
Concentrated Solar Power 112
CONCENTRATED SOLAR POWER 133
Conclusions 161
Conflict prevention 71
Construction and CO2e emissions 90
Copper 95, 108, 109, 140, 202, 203
Coral Reefs 51, 201
DAC 80
Deploying new capacity 75
Direct carbon capture 80
DROUGHTS 33
duty of the Federal Government 71
Economic Transformation 62
Electrical Output 86
ELECTRIC VEHICLES 140
Electrification. 76
eliminate pollution 74
energy-efficient 76
Energy Production 154, 191
Energy Projection 192

environmental destruction 65, 66, 111, 113, 126, 127, 134, 162, 163
family farms 78
Forests 189, 199, 201, 206
four-decade trend 64
Gallium 108, 109, 204
GEOTHERMAL 114, 129
German dilemma 165
GHG 8, 15, 16, 36, 37, 38, 41, 57, 67, 74, 92, 99, 100, 101, 103, 117, 122, 136, 147, 166, 204
GND 1, 2, 3, 4, 7, 8, 11, 12, 13, 15, 18, 27, 39, 41, 50, 55, 58, 60, 61, 62, 63, 64, 65, 66, 67, 68, 69, 70, 71, 74, 76, 77, 78, 79, 82, 83, 85, 125, 126, 127, 158, 172, 173, 191, 192
GND must be developed 83
Goals in A through E 73
government has to leverage 83
Green New Deal v, xi, 1, 2, 7, 11, 15, 71, 172, 199, 210
heat pump 76
heat stress 54
Human Activity 15
HURRICANES 24, 150
HYDROPOWER 116
Identify and create solutions 82
inadequate resources 64
INCREASE IN WILDFIRES 20
INCREASING WILDFIRES 149
Indium 108, 109, 204
industrial complex 172
infrastructure bill 74, 75
IPCC 2, 4, 7, 9, 11, 12, 13, 14, 16, 17, 33, 42, 47, 56, 58, 66, 121, 151, 153, 166, 171, 172, 174, 210
KEYSTONE PIPELINE 143
Lessen CO2e Emissions 156
life expectancy 63

long-term effects of pollution 79
Mass Migration 43
Migration 6, 148, 149, 202, 205
military- climate-change industrial complex 172
Mitigation 70
Mylar 108, 110
Nacelle 96, 203
Net-zero global emissions 58
OCEAN ENERGY 119
OTHER STUFF 36
Phosphorus 108
pollution 37, 53, 65, 66, 74, 79, 82, 115, 119, 129, 207
population 166
POPULATION GROWTH 147
Preindustrial Levels 41, 57
Professor Roger Pielke, Jr 166
public infrastructure 55
reducing the risks posed by climate impacts 74
reductions in GHG 57
renewable energy sources 75, 85, 175
renewables 1, 6, 7, 9, 10, 11, 12, 17, 33, 38, 44, 50, 53, 58, 62, 63, 65, 70, 72, 73, 74, 75, 76, 81, 85, 88, 105, 115, 118, 125, 126, 127, 136, 139, 142, 147, 148, 151, 152, 153, 154, 156, 157, 160, 161, 165, 166, 167, 169, 171, 173, 174, 175, 176, 210
Resolution 109 2, 11, 83
Restore and protect threatened ecosystems 82
SDG 6
Sea Level Rise 148
SEA LEVEL RISE 17
secure their list of things 72
Selenide 108, 109, 204

212

SEVERE SNOWSTORMS &
    BLIZZARDS 30
SEVERE STORMS 24
Silicon 108, 110
Silver 108, 110, 204
SNOWSTORMS AND
    BLIZZARDS 152
SOLAR 107, 130
Solar Panels 130
Sulfur 103, 110, 204
sustainable development goals 6
sustainable farming 78
THE BIG REVEAL 173
THE MONEY TRAIL 169

threat minimizers 69
threat multiplier 68
threat to the national security 66
Tin-coated copper 108, 109
TORNADOES 151
transportation system 79
upgrading infrastructure 74
wealth redistribution 173
What the Blazes is Going On? 1
wind turbines 176
Wind Turbines 87, 88, 89, 179, 183, 202, 203, 208
<u>WIND TURBINES</u> 85, 126
Zinc 108, 109, 204